ROADSIDE
Geology and Biology of
BAJA CALIFORNIA

by
John Minch
Edwin Minch
Jason Minch

194 Photographs
145 Line Drawings

John Minch and Associates, Inc.

Post Office Box 4244
Mission Viejo, CA 92690
(800) 367-2995, Fax (949) 367-0117

Based in part on *The Baja Highway* by John Minch and Thomas Leslie

First printing 1998

Printed in the United States of America
ISBN: 0-9631090-1-4

Library of Congress Cataloging in Publication Data
Minch, John A
Roadside Geology and Biology of Baja California / John A., Edwin S. Minch and Jason I. Minch
Derivitive work of: The Baja Highway, a Geology and Biology Field Guide for the Baja Traveler
Includes bibliographical references
ISBN: 0-9631090-1-4
1. Travel - Mexico - Baja California, 2. Geology - Mexico - Baja California, 2. Biology - Mexico - Baja California
I Minch, Edwin S. II Minch, Jason I. III Title
Library of Congress Number 98-065377

For copies of this book send $19.95 + 3.00 postage & handling payable to:
(California residents add California sales tax.)
John Minch and Associates, Inc.
P. O. Box 4244
Mission Viejo, CA 92690-4244
Phone (800) 367-2995, Fax (949) 367-0117

Inquire about multiple copy and
Distributor/Bookstore rates.

Production Notes: This manuscript was compiled entirely by CD-Rkive with the use of Macintosh Computers. The original manuscript was typed in Microsoft Word 6.0. It was then transcribed into Adobe Pagemaker 6.5. The figures were drawn and scanned with an HP Scanjet scanner into Adobe Photoshop 4.0. They were then captioned in Claris Macdraw Pro1.5v3 and transferred into Adobe Pagemaker 6.5. The photographs were scanned with a Microtek Scanmaker 35T into Adobe Photoshop 4.0 and transcribed into Adobe Pagemaker 6.5. The front and back color cover were scanned with the Microtek Scanmaker 35T into Adobe Photoshop 4.0. The manuscript was stored and delivered to the printer on CD-ROM by CD-Rkive. CD-Rkive directed and supervised the entire project and is available to assist others. Their offices can be reached through John Minch and Associates at (800) 367-2995.

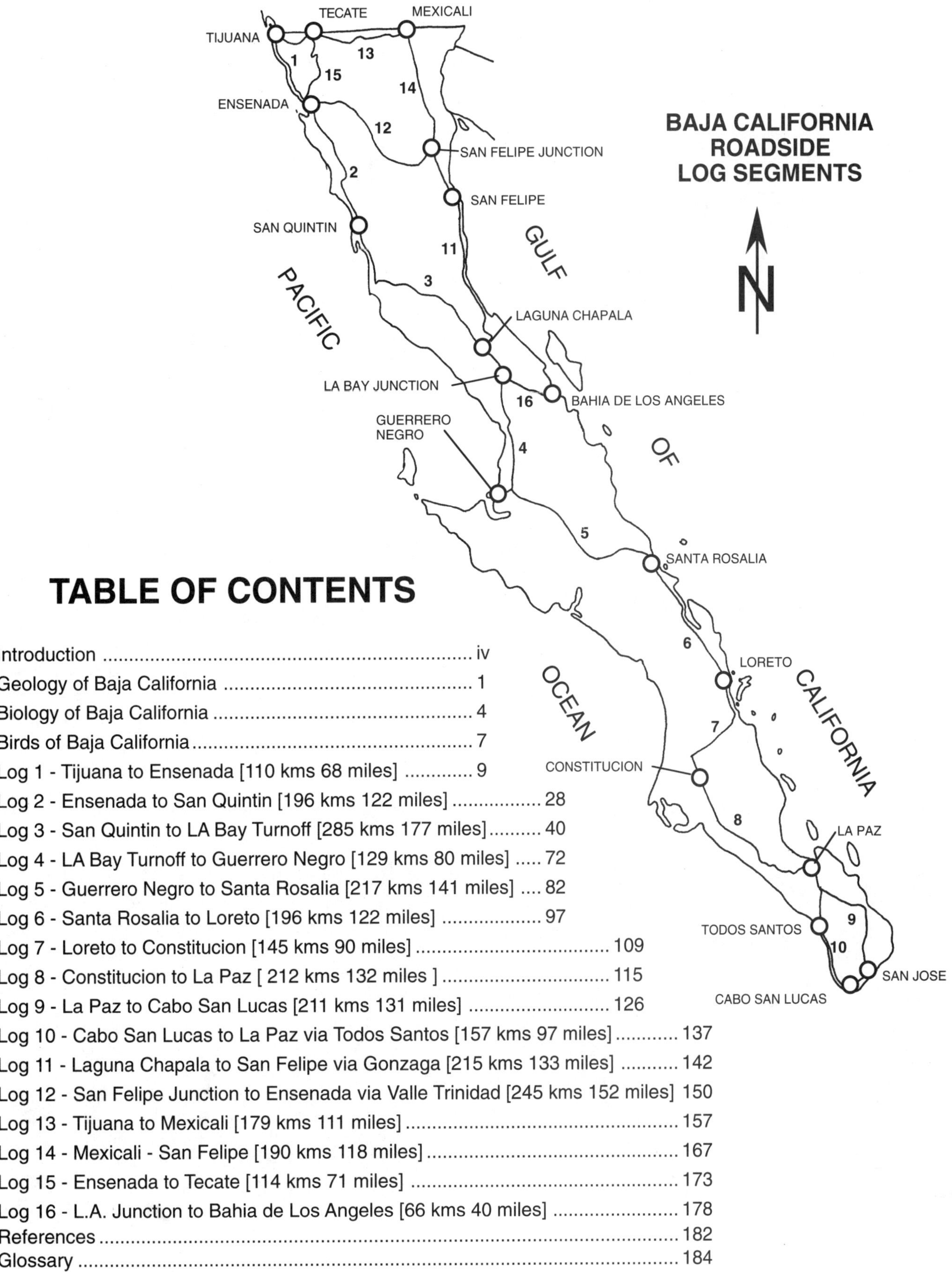

TABLE OF CONTENTS

Introduction iv
Geology of Baja California 1
Biology of Baja California 4
Birds of Baja California 7
Log 1 - Tijuana to Ensenada [110 kms 68 miles] 9
Log 2 - Ensenada to San Quintin [196 kms 122 miles] 28
Log 3 - San Quintin to LA Bay Turnoff [285 kms 177 miles] 40
Log 4 - LA Bay Turnoff to Guerrero Negro [129 kms 80 miles] 72
Log 5 - Guerrero Negro to Santa Rosalia [217 kms 141 miles] 82
Log 6 - Santa Rosalia to Loreto [196 kms 122 miles] 97
Log 7 - Loreto to Constitucion [145 kms 90 miles] 109
Log 8 - Constitucion to La Paz [212 kms 132 miles] 115
Log 9 - La Paz to Cabo San Lucas [211 kms 131 miles] 126
Log 10 - Cabo San Lucas to La Paz via Todos Santos [157 kms 97 miles] 137
Log 11 - Laguna Chapala to San Felipe via Gonzaga [215 kms 133 miles] 142
Log 12 - San Felipe Junction to Ensenada via Valle Trinidad [245 kms 152 miles] 150
Log 13 - Tijuana to Mexicali [179 kms 111 miles] 157
Log 14 - Mexicali - San Felipe [190 kms 118 miles] 167
Log 15 - Ensenada to Tecate [114 kms 71 miles] 173
Log 16 - L.A. Junction to Bahia de Los Angeles [66 kms 40 miles] 178
References 182
Glossary 184

ACKNOWLEDGMENTS

There are numerous people who contributed to the spirit of this book, both directly and indirectly. We wish to thank all of them for their measure of support. Particular thanks go to the help and support given by our mother and wife, Carol Minch and our family who put up with the endless hours of work on the manuscript and with the numerous trips into the heart of Baja to compile and check on the information in the logs. Jenifer King (an eighth cousin) made major contributions to the biology in rewriting difficult sections and providing important information on the birds, plants, and Phytogeographic Regions. We especially want to thank Thomas Leslie for his contribution to and work on the writing of the biology of an earlier book titled "The Baja Highway" (Minch and Leslie, 1991). That book provided part of the foundation on which this derivative work was built.

Jennifer Minch and Jenifer King assisted in the editing, research of new data, and the transcribing of the manuscript. Edwin Minch drew and drafted most of the illustrations. Jason Minch and Edwin Minch digitally scanned and imaged the figures and photographs. John's compadre Jorge Ledesma, Profesor de Geologica, Universidad Autonoma de Baja California en Ensenada, provided assistance and extensive critical technical review adding important information on the Roadside geology and biology of Baja.

Our special thanks also goes to Edwin Allison and Gordon Gastil who provided the spark of interest in Baja which led to a burning passion and a lifelong love affair with the peninsula. Their picture is at 3:171.5 where we gassed up near Catavina 30 years ago this year.

IN MEMORY OF EDWIN CHESTER ALLISON
ALISITOS LIMESTONE FROM PUNTA CHINA

HOW TO USE THIS BOOK

The logs in this book are keyed to the kilometer markings on the Highway. These markings have remained relatively constant for the last twenty years and should change little. The log segments correspond with the consecutive numbering system between major points. The kilometer markings, in each segment, increase in the State of Baja California from the border to Guerrero Negro and decrease to La Paz and Cabo San Lucas in the State of Baja California, Sur.

It is possible to pick up the book at any kilometer marking and read about the geology and biology along the highway. Some descriptions will be followed by a (*See* 4:32). This is referring you to a description in another location in the book. The 4:32 refers to Kilometer 32 of Log 4. The pictures and drawings flow with the text or are close to the text description.

At this Printing, the Laguna Chapala to Puertocitos stretch is still unpaved with poor to nonexistent kilometer markings. I have lost tires on each of my last three trips over that road.

GETTING STARTED

Most people enter Baja California at the San Ysidro border crossing south of San Diego. This is the busiest border crossing in the world. More than half of all of the people who leave the United States to visit a foreign country pass through Tijuana.

Since you will probably be following Log 1 to Ensenada, the following directions will be very useful.

For the most direct route to the toll road and points south, get into the third lane from the center of the road as you drive under a large gray arch (a pedestrian overpass) and continue across a bridge over the Tijuana river. Do not change lanes until you are heading north down the Tijuana River. Shortly after you cross over the river, you will come to a division in the road. The left two lanes will take you into the central district of the City of Tijuana. **Stay in the lane to Ensenada!** After the split you will be in the far left lane of two lanes. If you stayed in the third lane, as directed above, you will continue around a 270 degree circle, under a bridge, and head northwest on Highway 1D, paralleling the Tijuana river. The road bends to the left and parallels the International Border.

GEOLOGY OF BAJA CALIFORNIA

Geomorphic Provinces

Baja California can be divided into 5 geomorphic provinces (based on geologic landforms). They are: 1) the tilted fault blocks of the *Peninsular Ranges Batholith*, 2) the broad flat plains of the *Cretaceous Geosyncline*, 3) the isolated mountains of the *Coast Ranges*, 4) the fault block mountains and alluviated valleys of the *Basin and Ranges*, and 5) the plateaus of the *Volcanic Tablelands.*

The first three provinces are part of the Cretaceous collision of the North American and the Pacific Plate.

1) The ***Peninsular Ranges Batholith*** is represented by the main granitic ranges of the State of Baja California (such as the Sierra Juarez and the Sierra San Pedro Martir) and by the Cape region mountains (Sierra La Laguna). This province represents the Peninsula Ranges batholith in Baja California. The main rock types are granitic (granite, granodiorite, tonalite, gabbro), metamorphic (schist, gneiss, marble), and metavolcanic (metamorphosed volcanic rocks such as andesites, sandstones, and breccias).

2) The ***Cretaceous Geosyncline*** is largely offshore in Northern Baja (Norte) with only a narrow fringe exposed on land (widest near El Rosario). In Baja California, Sur it underlies nearly all of the state outside of the Cape region. It is generally represented by broad flat areas (Vizcaino and Magdalena Plains) or hidden under the Volcanic Tableland. This province is similar to the Great Valley of California with up to thirty thousand feet or more of sandstone, shale, and conglomerate deposited in a basin in a broad belt in the subduction zone.

3) The ***Coast Ranges*** is not seen along the highways in Baja California. In Baja California (Norte) it is offshore. In the State of Baja California, Sur it is isolated in the mountains of Cedros Island, the Vizcaino Peninsula and the Magdalena Bay islands. Debris from these rocks can be seen in the Rosarito Beach Formation near Tijuana. This province is the Coast Range Province of California. It represents the scrapings from the sea floor of the Pacific plate and mantle rocks (serpentine) which have been metamorphosed and brought back to the surface.

The last two provinces are directly or indirectly related to the opening of the Gulf of California:

4) The ***Basin Ranges*** forms the Gulf of California area east of the main Gulf escarpment. Nearly all of the blocky mountain ranges and alluviated valleys of Baja California belong to this province. This province cuts across older provinces. It is based on the presence of block faulting related to the opening of the Gulf of California.

5) The ***Volcanic Tableland*** forms a broad plateau and mesas, north of the Cape region, in Baja California Sur. These rocks were formed as a result of the passing of the East Pacific rise under the continent and the beginning of the opening of the Gulf of California. They represent extensive outpourings of volcanic material (tuff andesites, basalts, breccias and the reworking of this material into volcanic sandstone and conglomerates).

Erosion Surfaces

When the dinosaurs became extinct, Baja California was a relatively flat place with flat to gently rolling topography stretching to the east across present day Sonora. Many remnants of these surfaces are exposed along the highway and dominate the landscapes over which you will drive. Streams are cutting into these surfaces and rapidly destroying them along their edges, however, major pieces of them still exist today (3:82, 3:115, 12:110, 13:109, 15:40).

Tilted Fault Blocks

The East Pacific rise passed under the continent and began to open the Gulf of California. After some resistance, the continent split, the Gulf opened, and the eastern side of Baja California was arched upward forming the tilted fault blocks of the main ranges. The enormity of these fault blocks is realized when you gently climb for tens of miles along the western side and then steeply descend the eastern escarpments (5:35, 8:34, 13:65).

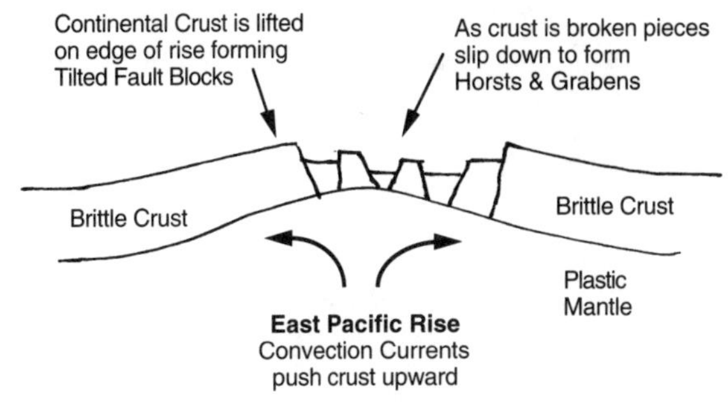

VULTURES CUEVA PINTADA

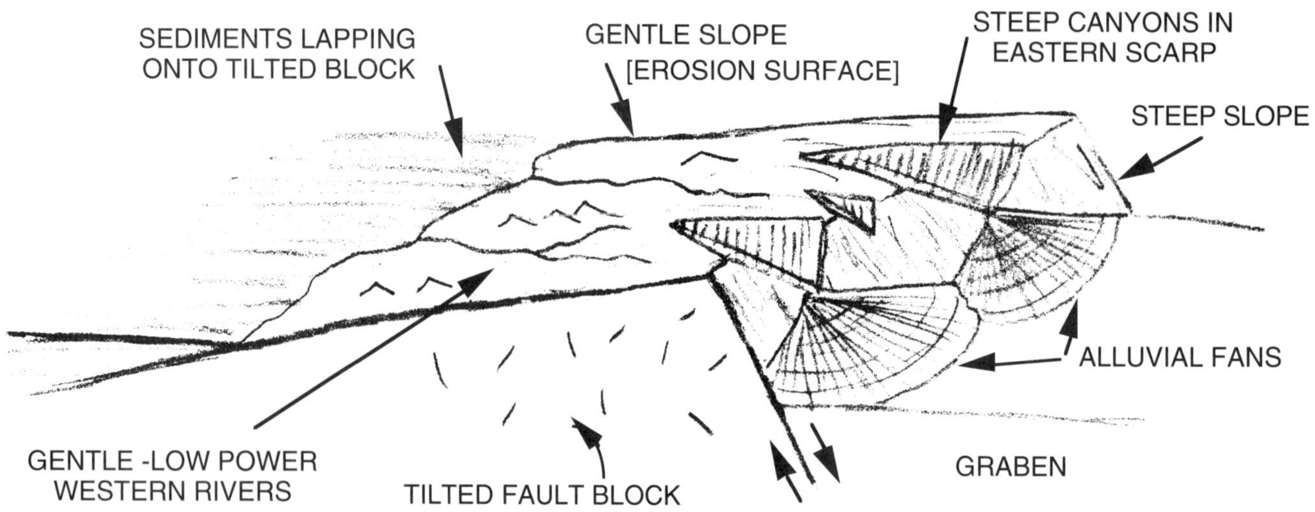

STYLIZED CROSS-SECTION OF TILTED FAULT BLOCK

Baja Geologic History

Plate Tectonics

The crust of the earth is a series of semirigid plates, moving about relative to each other, and "floating" on the mantle. When the plates pull apart they form Ridges or Rises in Oceanic Crust and rifts [Gulf of California] in Continental Crust. When they push together they form Subduction Zones and Island Arcs. When they slide past each other they form Lateral [strike-slip] Faults. It is the interaction of the plates, with each other, that is producing the major features on the earth's crust. The vast majority of this motion is very slow and literally occurring before our very eyes on a daily basis. An example is Cajon Pass in southern California. A surveyed point on the railroad through the pass is rising at a steady rate of 16"/100 years. That's 16,000 feet in a million years (See 13:64)

Paleozoic - The Quiet Time

The Pre-Mesozoic history of Baja California is obscure, very fragmental, and not well documented. We know that what would later become the North American and Pacific Plates were together as part of a larger plate which was moving eastward closing the Proto-Atlantic Ocean on its way to a collision with Europe and later, Africa. The west coast was in the middle of this plate like the present eastern coast of North America is today. This stable edge of the continent was receiving clastic sediments from the continental landmass while limestones were being deposited in shallow continental shelf and slope areas. This added a wedge of sediments to the continent. The end of the Paleozoic was marked by the formation of the supercontinent of Pangaea which included all of the continents. In Baja metamorphosed remnants of Lower Paleozoic carbonates, shales and sandstones occur in at least one place while Upper Paleozoic rocks have been identified in a number of isolated and scattered areas. Some of the metamorphic gneisses and schists such as the Julian Schist are most likely of Paleozoic age.

Mesozoic - The Big Squeeze

At the beginning of the Mesozoic the motions changed. As the supercontinent broke up and the Atlantic opened, North America began to move westward pushing against the thin oceanic crust of the Pacific part of the plate. The thin edge of the continent buckled and the heavier oceanic crust was forced under the lighter continental crust forming a Subduction Zone and Island Arc with the accompanying volcanoes. Parts of Baja may have been brought from other areas as small land masses on the East Pacific Plate. However, the majority of the intrusive igneous and metavolcanic rocks in Baja were formed in this subduction zone.

As the Pacific plate was pushed under the continent, the friction caused melting and the formation of magmas. Oceanic crust was continually pushed into and added to the continent in this subduction zone resulting in uplift and the raising of the magmas closer to the surface. Some of them cooled miles below the surface to form the granitic rocks of the Peninsular Ranges Batholith with its accompanying metamorphic rocks. Some of the magma spewed out on the surface and formed volcanoes with the associated volcanic derived sedimentary basins.

Continual subduction and burial of these oceanic crustal rocks caused low grade metamorphism and formed the Alisitos Formation and other metavolcanic rocks. As the mountain mass was raising, erosion carried much of the debris westward into the trench basin forming a large Mesozoic sedimentary basin (geosyncline) offshore. The subduction continued for a hundred million years resulting in the formation of more plutonic rocks after the earlier plutonic rocks were uplifted to the surface and the eroded sediments deposited offshore in the geosynclinal basin and on top of earlier granitic rocks. The fringes of this basin are exposed as the Rosario Formation along the coastline in northern Baja and as a 30,000+ foot thick geosynclinal basin under the majority of southern Baja. The overriding of the East Pacific Rise by the North American plate resulted in the end of the subduction.

Cenozoic - The Big Split and Rip-off

In the Early Cenozoic, the peninsula was again a relatively quiet place. The peninsular sierras were worn down and a gently rolling erosion surface was developed on the exposed batholithic rocks. This surface stretched to the east well into Arizona and Sonora. Major rivers, bearing gravel, flowed across the area from central Arizona to the Pacific.

The North American Plate overrode the East Pacific Rise and began the great rip-off. Coastal California and Baja California began to slide northward along strike slip faults like the legendary San Andreas Fault in California and the San Miguel Fault, Agua Blanca Fault, Vizcaino Fault, Magdalena Bay Fault and others in Baja California.

The middle Cenozoic opening of the rift, later to become the Gulf of California, took tens of millions of years. Great sheets of lava and pyroclastic rocks with accompanying volcani-clastic sediments were spread over large areas of the peninsula during the Miocene and Pliocene and shallow Miocene seas spread across low areas of the southern part of the peninsula to fill tectonic basins opening in the Proto-Gulf area. The present shape and form of Baja California was developed in the last 5-10 million years as the continent finally yielded to the stretching and opened successive areas of the Gulf, finally, opening the mouth about 5 million years ago. The splitting of the continent tilted the peninsula westward forming the asymmetric tilted fault blocks of the main ranges of the Sierra Juarez, Sierra San Pedro Martir, Sierra la Giganta and uplifting other ranges such as the Sierra la Asamblea and Sierra la Victoria.

STYLIZED GEOLOGIC CROSS-SECTION OF BAJA CALIFORNIA

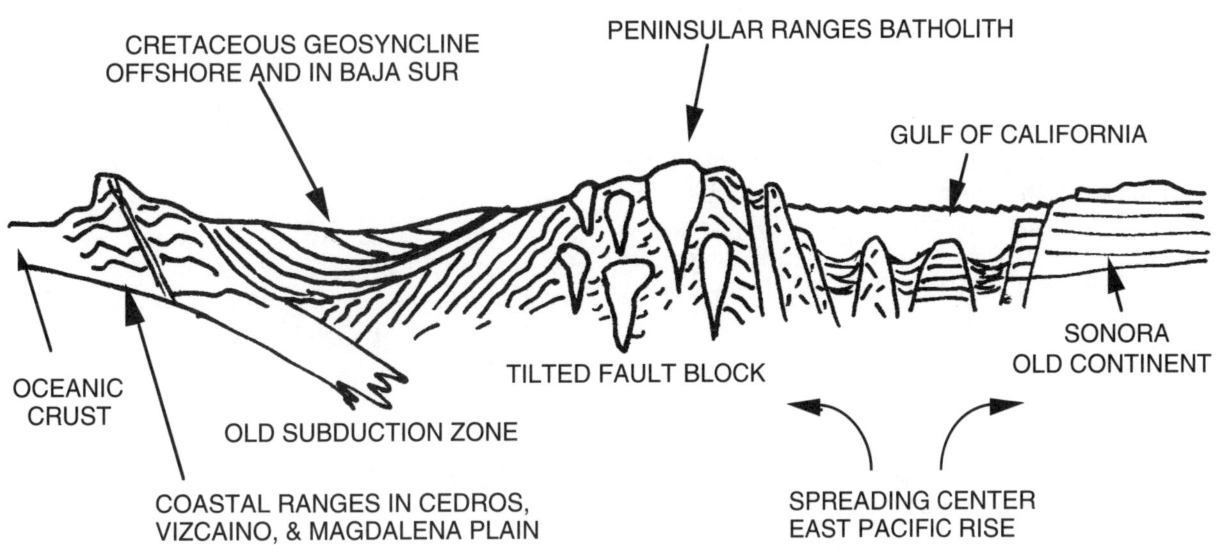

BIOLOGY OF BAJA CALIFORNIA

BOTANICAL WORK ON THE PENINSULA

Much work has been done on the higher vascular plant species of Baja. Plant collecting was conducted in Baja as early as 1841 by the naturalist I. G. Voznesenskii. In this century several excellent botanical collections, botanical accounts, and published references have become available which detail the floristics of the vascular plants of Baja for interested travelers, students, naturalists, and serious botanists.

Those references which are particularly noteworthy and useful are by Roberts (1989), Standley (1920&26), Shreve and Wiggins (1964), Felger and Lowe (1976), Axelrod (1979), Wiggins (1980) and Copyle, 1990.

PHYTOGEOGRAPHIC AREAS

Traveling along the peninsula, the traveler will notice changes in vegetation. The familiar plants of southern California's chaparral and mountains are succeeded by unfamiliar Desert Region plants. Farther south the equally strange species of the Cape Region appear. These plant ranges are divided into regions, or phytogeographic areas.

Three major phytogeographic regions of the peninsula are recognized by Shreve and Wiggins (1964), Wiggins (1980), and Roberts (1989).

The **CALIFORNIAN REGION** is an extension of the southern California mountain ranges and their Pacific drainages south to the Sierra San Pedro Martir and some of the mountains in the middle of the peninsula. Vegetative communities of the Californian Region encompass Coniferous Forest, Juniper-Pinon Woodland, Chaparral, Coastal Sage Scrub, and Riparian areas.

The Californian Region plant communities and climatic conditions, similar to those of southern California, are found south to San Quintin. Coniferous forests are found in the Sierra Juarez and the Sierra San Pedro Martir. These blend into Pine-Oak Woodlands which are replaced by Chaparral and, near the Pacific, by Coastal Sage Scrub.

Coniferous Forests. Coniferous forests are found along the crests of the Sierra Juarez and the Sierra San Pedro Martir.

Common species are: Jeffrey Pine, Incense Cedar, Sugar Pine, Lodgepole Pine, Pinon Pine, and White Fir. Other coniferous forest species are Ceanothus, Manzanita, Quaking Aspen, and Canyon Live Oak.

Pine-Juniper-Oak Woodland. Pine-juniper-oak woodland is a transition between the pine forests and the chaparral.

Common species are Pinon Pine, Coast Live and Scrub Oak, Maguey, Manzanita, Yucca, Penstemon, Rabbitbrush, Sage, Barrel Cactus, and Lilac.

Chaparral. The chaparral covers areas of the western slopes of the mountains south to San Quintin. Chaparral can range from of nearly pure Chamise (Chamisal), to one of the most diverse of plant communities.

Chaparral species include Broom Baccharis, Bush Monkey Flower, Ceanothus, Live Oak, Scrub Oak, Flat Top Buckwheat, Indian Paint Brush, Laurel Sumac, Lemonade Berry, Manzanita, Mexican Elderberry, Yucca, Sage, Toyon, and Lilac.

Coastal Sage Scrub. Coastal Sage Scrub is similar to, but, distinct from Chaparral, with a more open appearance, and more Barrel and Hedgehog Cacti. The shrubs are less evergreen, rigid, and woody, with thinner, softer leaves. Many of the species are partially deciduous, dying back during drought periods.

Coastal Sage Scrub species include: Agave, Barrel Cactus, Sagebrush, Broom Baccharis, Chamise, Cholla, Ceanothus, Dudleya, Fishhook Cactus, Flat-Top Buckwheat, Hedgehog Cactus, Jojoba, Laurel Sumac, Lemonade Berry, Mormon Tea, Mountain Mahogany, Pitaya Agria, Prickly Pear, and Toyon.

Riparian Areas. Riparian Woodlands are vegetational communities which grow along streams and other drainage ways. Since the climatic regime over much of Baja is an arid one, the local occurrence of standing or running water has a dramatic influence on the composition and quantity of vegetation. Localities with permanent standing or running water are generally bordered by deciduous trees, shrubs, and herbs which only grow on the banks of such water courses. Where river valleys are broad, the Riparian Woodland is correspondingly broad; at higher elevations where the water courses are narrow and the stream banks are steep, Riparian Woodlands may form a very narrow strip which may be only a few meters wide.

In the arroyos of the Californian Region, the canopy consists dominantly of Willows with Cottonwoods, Oaks, and Sycamores. The understory is Baccharis, Chamise, Lilac, Scrub Oak, Wild Rose, and Willow.

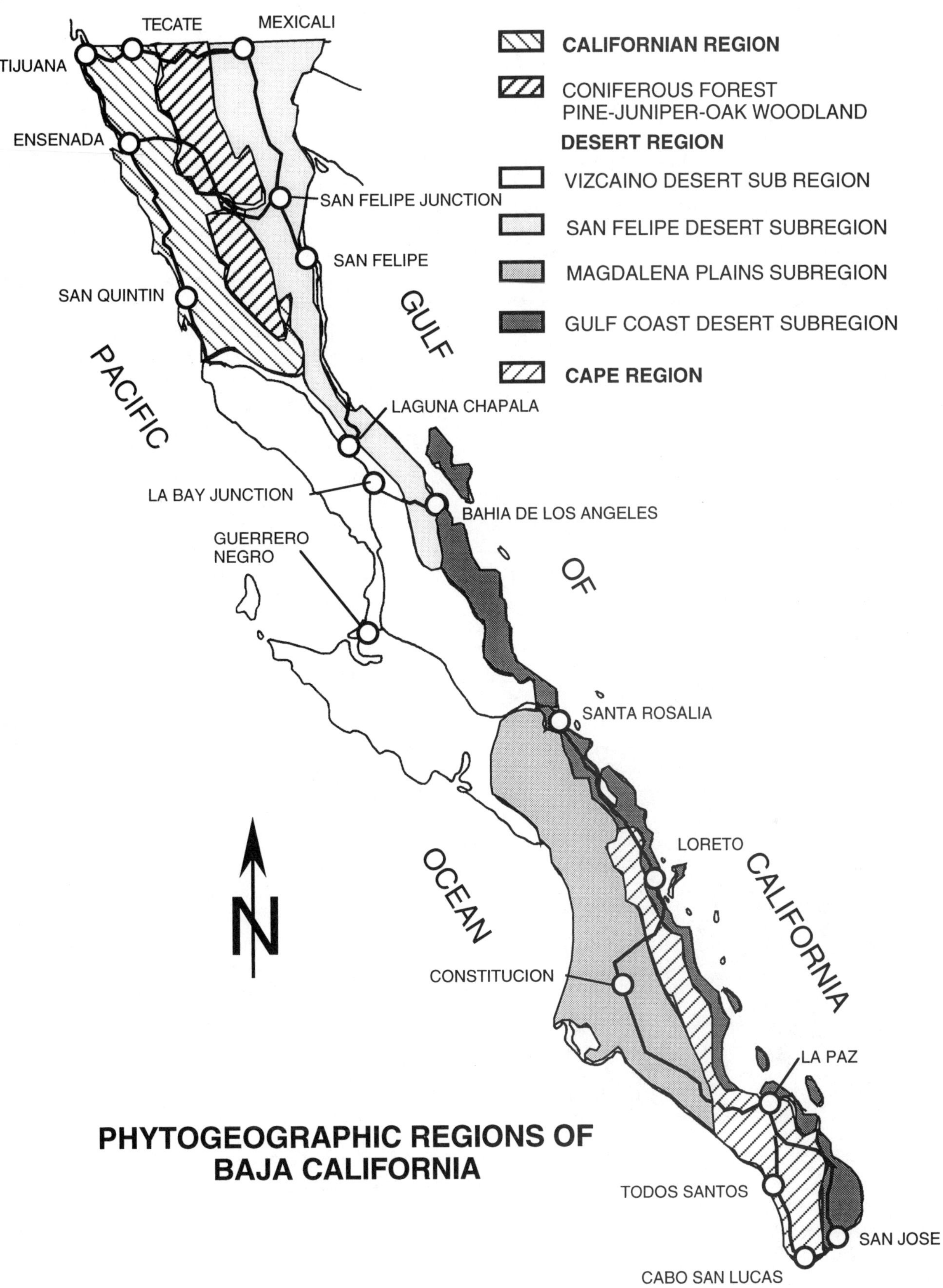

PHYTOGEOGRAPHIC REGIONS OF BAJA CALIFORNIA

The **DESERT REGION** comprises the majority of the peninsula, including a narrow strip along the east coast, and the central and southern regions of the peninsula. The Desert Region is divided into subregions: San Felipe, Gulf Coast, Vizcaino, and Magdalena Deserts.

The Desert Region is south and east of the Californian Region, climate and vegetation resemble that of the Sonoran Desert, which covers much of the Mexican states of Sonora and Sinaloa and extends into southeastern California and southern Arizona.

Two-thirds of Baja California and all the islands of the Gulf have Sonoran Desert vegetation.

Isolation of the Peninsula has resulted in the development of unusual plants such as the Cardon, Cirio, Elephant Tree, Ocotillo, and Candelillo.

Common plants that occur in one or more areas of the desert region include: Biznaga, California Fan Palm, Candelillo, Cardon, Cholla, Cirio, Datilillo, Desert Mistletoe, Devil's Claw, Dodder, Garambullo, Cactus, Lomboy, Maguey, Mesquite, Ocotillo, Palo Adan, Palo Verde, Pitaya Agria, Pitaya Dulce, and Torote.

CORDON AND MOON AT CATAVINA

The Desert Region is commonly divided into four subregions each with its own dominant species. Many of them occur in more than one subregion. The San Felipe Desert, Gulf Coast Desert, Vizcaino Desert, and Magdalena Plains are all part of the Desert Region.

San Felipe Desert Subregion. The vegetation is very sparse in most areas. Ninety percent of the vegetation in this region is Bursage and Creosote Bush. Other common plants are Brittlebush, Cardon, Cholla, Desert Willow, Indigo Bush, Ironwood, Mesquite, Ocotillo, Palo Verde, Pitaya Agria, Smoke Tree, Torote, and Yucca.

Gulf Coast Desert Subregion. This region is characterized by trees with swollen trunks including Lomboy, and Torote. Brittlebush, Bursage, Cardon, Cholla, Cirio, Creosote Bush, Ironwood, Ocotillo, and Palo Blanco are significant parts of the flora.

Vizcaino Desert Subregion. Most of the dominant plant are succulents or cacti. Though there are Mesquite in many arroyos, tree are scarce. Common plants are Agave, Ball Moss, Burbush, Bursage, Cholla, Cirio, Copalquin, Creosote Bush, Ocotillo, Opuntia, Palo Adan, Pitaya Agria, and Yucca.

Magdalena Plains Subregion. The Magdalena Plains have an abundance of cacti such as Cardon, Cholla, Garambullo, and Pitaya Agria. Mesquite, Palo Adan, Palo Blanco, Palo Verde, and San Miguel are found in the foothills, and Mesquite groves grow in the arroyos. The "fog desert" of this area encourages lichens and Gallitos on the shrubs and cacti. Rocella *(Ramalina* species) is the most noticeable lichen south of Punta Prieta. Other common plants are Datilillo, Lomboy, Maguey, Pitaya Agria, Pitaya Dulce, and Torote.

The **CAPE REGION** includes the Sierra de la Laguna and Sierra de la Giganta and the area south of La Paz. Vegetative communities are Oak-Pinon Woodland and Arid Tropical Forest.

The Cape Region is classified as arid tropical forest. Oak-Pinon Woodlands are at higher elevations and a semi-deciduous forest, prominent in leguminous trees and shrubs occupies lower elevations.

Cape Oak-Pinon Woodland. The Oak-Pinon Woodland community is found at higher elevations in the granitic Sierra de La Laguna south of La Paz.

There are extensive Mexican Pinon Pine stands in addition to Palmita, Madrono, and Oaks. Some Oaks are found at higher elevations in the Giganta.

Cape Arid Tropical Forest. The arid tropical forest has shrubs and trees of different height and branching character spreading and intermingling with each other making the region look like an "impoverished tropical jungle," as described by Shreve.

Dominant trees include: Mauto, Mesquite, and Palo Verde. On lower elevations Cardon, Coral Tree, Palo Blanco, Palo de Arco, San Miguel, and Yuca occur.

BIRDS OF BAJA CALIFORNIA

Birds are seen during the day along the Peninsular Highway. A traveler taking a walk into the different plant communities adjacent to the highway will be rewarded with views of a variety of birds. Although there are some areas that are inhabited by more species than others, some Baja travelers think birding is wonderful throughout the entire peninsula.

ISLA PAJAROS IN BAHIA MAGDALENA

The birds of Baja generally exhibit regional habitat preferences and so we find, as with plants, that specific species of birds tend to be found in specific geographical locations. The birds seen along the highway change locally with the season, geography, latitude, temperature, and weather.

Binoculars are useful for birdwalks in Baja. A little preparation, involving looking at pictures of birds commonly seen in Mexico will help identify the birds on or adjacent to the Peninsular Highway.

Good sources of pictures and information can be found in Robbins, Brunn, and Zim (1983), Wilbur (1987), Scott (1992), Radamaker (1995), and Peterson's Field Guide to Western Birds.

In his ABA field check list, Radamaker (1995) names 450 species of birds recorded from Baja. The birds of Baja's arid deserts, however, are poorly known, species lists are incomplete, and very few studies concerning the ecology and densities of breeding pairs per acre have been conducted and reported. In the more arid, desert regions of Baja there are fewer birds per acre than in the more dense California region and those commonly seen are generally drab colored for camouflage and thermoregulation. Most of Baja's plants lose their leaves during the drier parts of the year and consequently birds have little cover to hide behind so they are more visible to birdwatchers.

Four species in particular distinguish the avifauna of the peninsula deserts. They are Scrub Jay, California Quail, Xantus' Hummingbird, and Gray Thrasher. The last two are both endemics of the southern two-thirds of the peninsula and are found in both the desert areas and the Cape Region.

Since most of the birds commonly seen along the Peninsular Highway are also commonly seen throughout both of the Americas, it is difficult to typify the various regions of Baja on the basis of its avifauna. However, an attempt has been made in this guide to list those species most commonly seen along the Peninsular Highway as the traveler passes through the three main phytogeographic regions and the eight subregions of the peninsula. Some of Baja's bird species are cosmopolitan over the entire peninsula and so are listed in more than one phytogeographic region.

Several cosmopolitan bird species occur from the International Border to the Cape and along both coasts and on the Gulf islands, in suitable habitats. They are: Costa's Hummingbird, Gila Woodpecker, California Quail, Raven, House Finch, Ladder-backed Woodpecker, Red-tailed Hawk, American Kestrel, Ash-throated Flycatcher, Horned Lark, Verdin, Cactus Wren, Rock Wren, Northern Mockingbird, Blue-gray Gnatcatcher, Black-tailed Gnatcher, Hooded Oriole, Least Tern, and Harris' Hawk.

The following information is intended as a brief introduction presenting in a general fashion the birds commonly seen along the highway in each phytogeographic region, along Baja's shoreline, on gulf islands, and Baja's raptors and migratory birds.

CALIFORNIAN PHYTOGEOGRAPHIC REGION:

The bird species of the Pacific coast of Northern Baja are the same as those found in the Californian Region of California. The dominant species of the Region are the California Thrasher, Black-shouldered Kite, Anna's Hummingbird, Black-chinned Hummingbird, Red-shafted Flicker, Nuttal's Woodpecker, Scrub Jay, Plain Titmouse, Wrentit, Bell's Vireo, Lawrence's Goldfinch, and Brown Towhee (See 1:35.5)

DESERT PHYTOGEOGRAPHIC REGION:

The dominant bird species of the Desert Region including its four subareas are, Gambel's Quail, Vermilion Flycatcher, LeConte's Thrasher, Crissal Thrasher, Abert's Towhee, Gray Thrasher, and Sage Sparrow (See 5:118, 7:111, 11:16, and 14:165).

CAPE REGION:
The most commonly encountered species of the Cape Region avifauna (estimated to be 250 species) are the American Robin, Crested Caracara, Northern Cardinal, Pyrrhuloxia, Yellow-billed Cuckoo, Bell's Vireo, Yellow-eyed Junco, Varied Bunting, Common Ground Dove, White-winged Dove, Xantus' Hummingbird, Blue-gray Gnatcatcher, Yellow Warbler, Belding's Yellowthroat. Many of the more commonly encountered Cape species are also found in the Desert of the American Southwest (See 8:2).

SHORELINE AND GULF ISLAND BIRDS:
The most common shore birds are the American Oystercatcher, Osprey, Gulls, Terns, Pelicans, Egrets, Magnificent Frigate-birds, Blue and Brown-footed Boobies, and the Great Blue Heron. These species are often seen on both the Pacific and Gulf coasts of the peninsula and on virtually every Gulf island. Other birds that are seen on the shoreline include: sandpipers, curlews, plovers, avocets, stilts, cormorants, and grebes. Although there are several peninsular endemics, there are no endemic birds on the islands in the Gulf of California. There are other species of shore birds but they are more patchily distributed in mangrove-fringed beaches and coves on the larger southern islands and along the southern peninsular Gulf coast.

RAPTORS:
Raptors enjoy a wide distribution throughout the peninsula because they lack a habitat preference and are capable of traveling over long distances. The species of raptors most widely distributed in Baja include: Red-tailed Hawks (14:165), Turkey Vultures (8:36), American Kestrels (1:45), Peregrine Falcon (8:25), Northern Harrier (12:145), Red Shoulder Hawks (3:147), Ospreys (4:128), and Caracara (9:176). Most of these predatory or scavenging birds are year round residents of the entire peninsula.

OSPREY IN FLIGHT AT GUERRERO NEGRO

MIGRATORY BIRDS:
Baja lies on the Pacific Flyway, the pathway of many bird species that breed in the summer in western North America and winter in Baja or further south. As a consequence, a large number of birds, such as Mallards, Brants, teals, and geese, are seasonally seen in Baja. San Quintin Bay is the winter home for the Black Brant and other waterfowl. In the early morning and early evening in Magdalena Bay there are thousands of birds flying by to their evening roosts. Low sandy Isla Pajaros looks like there are several birds in every square meter.

CALIFORNIA BROWN PELICAN

PELICAN CHICKS ON ISLA SAN LUIS

NOTE: In each Phytogeographic Region, a list of commonly encountered birds is listed (Tables 1:35.5, 5:118, 7:111, 8:2, 11:16, and 14:165). At specific points along the Peninsular Highway, where a particular species of bird has been commonly seen, a brief natural history discussion is presented the bird, along with a sketch of the bird illustrating distinctive identifying characteristics. The birds Mexican name is included if known.

Log 1 - Tijuana to Ensenada [110 kms = 68 miles]

The Highway climbs west out of the Tijuana River Valley through steep roadcuts in the terraces developed on the conglomerates and sandstones of the ancestral Tijuana River delta. It then turns south on the slopes between the high mesas and the low terrace of the Playas de Tijuana to follow the rugged basalt sea cliffs along the elevated Pliocene shoreline with steep rugged canyons, then drops onto a narrow, late Pleistocene terrace cut into the basalt cliffs and finally out into an area of gently rolling basalt and tuff hills. The Highway skirts Rosarito on the Pleistocene terrace with views of volcanic capped Mesa Redonda and Cerro Coronel and the rolling volcanic hills, then continues past Punta Descanso where the rolling hills are largely underlain by the marine sedimentary rocks of the Rosario Formation.

By Descanso the Highway is alternately following the narrow late Pleistocene terrace or climbing onto the numerous slump blocks of volcanic rock from the high Miocene basalt capped mesas which form the back drop for the seacliffs. The Highway crosses the steep sided canyon of the Guadalupe River and travels along a wide segment of the Pleistocene terrace at La Salina before climbing back onto the mesas and slump blocks with the high mesas of resistant basalt overlying the soft marine sedimentary rocks. At San Miguel the Highway follows the Pleistocene terrace past hills of marine sedimentary rocks and then metavolcanic rocks to Ensenada.

Kilometers:

0 International Border: As you cross the border, you will briefly pass through Mexican customs.

The highway passes under an arch and makes a sweeping 270° bend to the left to follow the Tijuana River northward toward the border (*See* the Introduction for directions).

2 The highway bends to the left and parallels the International Border ("Tortilla Curtain" as it is sometimes called) on the right (north).

FLOOD CONTROL ON THE TIJUANA RIVER: The dissipater system of flood control on the Tijuana River can be seen to the northeast. Flooding has been a continuing threat in the Tijuana River valley. About 73% of the 1,731 square miles of drainage basin is in Mexico. The river flows across the International Border 9.4 Kilometers from its mouth at the Tijuana estuary in Imperial Beach, California. Although about 72% of the basin is controlled by reservoirs, even moderate rainfalls have caused flooding in the lower reaches of the river. Construction began a joint flood control project on the Mexican side in 1972, and by 1976 the first phase of the project was completed. This consists of a concrete channel 80 m. wide extending 4 Kms. to the southwest from the border.

It was feared that a similar concrete channel on the U.S. side of the border would cause major environmental damage to the fragile and endangered habitat of the Tijuana estuary, so the portion in the U.S. has been constructed as a dissipater system. (The system includes flood plain zoning throughout most of the valley; levees for the low-lying areas along the Mexican border and Interstate Highway 5, a 1.1 Kilometer low energy dissipater structure at the border to slow down and spread out major runoffs entering the U.S. from Mexico, and a low flow channel with a desilting basin to handle minor floods.) A land use plan was also adopted which provides for an enlarged state park in the west end of the valley and for agriculture in most of the east end.

In the latter months of 1984 serious flooding and overflowing of the Tijuana sewage system forced closures of many of the beaches in southern San Diego County. The dissipater system functioned as planned, but the sewage input was not considered in the original plan. The Mexican and U.S. governments have constructed a sewage treatment plant on the U.S. side of the border to remedy this problem and prevent its recurrence.

4 The highway begins to climb steeply into the hills to the west, leaving the flat Tijuana River Valley behind. As the highway ascends it passes through a series of road cuts which expose conglomerates and sandstones which represent a fluvial part of the Pliocene age San Diego Formation. These delta sands and gravels were deposited into the San Diego embayment by the ancestral Tijuana River during the Pliocene and Pleistocene. Notice the steep almost vertical slopes of the road cut to the left. The material of this road cut is stable enough to resist gravity and stands in almost vertical slopes. However, rain water is making serious cuts in the slopes.

6.3 Turn right to stay on Highway 1D. The Tijuana River and the Border are on the right.

AN UNEXPECTED DAM: During the winter of 1965-66 an unexpectedly violent storm occurred and torrential rain produced a raging river in this usually dry canyon bottom. The waters of the swollen raging river were loaded with mud and debris that quickly clogged the small drainage culverts which pass under the highway. As a result, the highway fill acted as a dam and 2/3 of the canyon to the south of the Highway was filled. The "lake" threatened to flood the adjacent Imperial Beach and Rheem field area in the U.S.A. Several weeks were anxiously spent pumping water out of the "lake" until workers were able to free the clogged culverts.

9 Highway 1D curves southward towards Ensenada. A turnoff heading westward from the highway is the only access to the "Bullring by the Sea" and the housing development of Playas (beaches) de Tijuana. The Playas de Tijuana are developed on a 55,000 year old Pleistocene terrace equivalent to the low Nestor Terrace in the San Diego area.

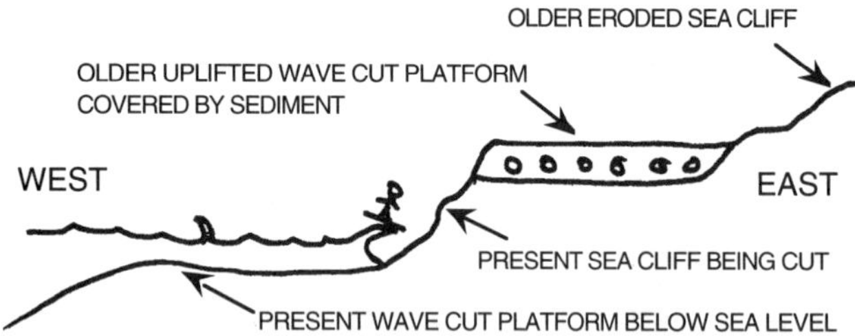

SEA CLIFFS, WAVE CUT BENCHES, MARINE TERRACES, AND BEACHES — LANDFORMS OF MARINE EROSION: As waves pound against shorelines, their impact (6,000 lbs./sq.in.) erodes uplifted **sea cliffs** back until a **wave-cut bench** or platform develops at their base. As erosion continues, the wave-cut bench widens until its width absorbs most of the wave energy and a **beach** forms along the now low-energy shoreline. The sea cliff diminishes over time due to weathering, erosion, and landslides. In regions of coastal uplift, like Baja, these wave-cut benches, called **marine terraces,** are raised above sea level. Generally, the highway will traverse these marine terraces whenever it is near the Pacific coastline.

Along the beach in the Playas de Tijuana there has been serious sea cliff erosion in the soft terrace sediments which resulted in the disappearance of several city blocks from the coastline. This coastline erosion may destroy more of the housing development presently located there.

There is a good view to the west of Islas Coronados. The toll road (Highway 1D) skirts around the eastern border of the Playas de Tijuana area and continues southward to Ensenada through marine conglomerates and sandstones of Pliocene and Pleistocene age which represent the edge of the ancient Tijuana Delta. Deltas form when a stream carrying suspended sediments enters standing water, where its velocity is checked, and the suspended sediments drop out forming a fan-shaped or triangular delta similar to the Greek letter "delta".

10 There are three "Casetas de Cobro" (hut of the fees) toll gates between Tijuana and Ensenada. The fee fluctuates. They can be paid in either Mexican or U.S. Currency.

12 If you need help on the highway to Ensenada, there are 29 solar-powered phones, located, every two or three Kms, on 7 meter high light-blue poles, and marked by large, highly visible blue and white rectangular signs bearing the internationally recognized distress letters "S.O.S".

As the highway heads southward to the Cape region, a number of gradual vegetational changes are obvious. Changes in regional and local vegetational associations will be discussed as they are encountered.

The **CALIFORNIAN PHYTOGEOGRAPHIC REGION** dominates northern Baja. In this area there are five vegetational groups which are the Coniferous Forest, Pinon Juniper Woodland, Chaparral, Coastal Sage Scrub, and Riparian areas. Each area is typically characterized and recognized by its association of dominant plant species known as indicator species.

INDICATOR SPECIES: Every plant is a product of the conditions under which it grows and is a measure of its environment. Certain plant species may be restricted exclusively to a single soil type or may occur in only a certain climatic regime and are referred to as indicator species. For example, in Baja, Creosote Bush grows only in arid deserts, Smoke Trees grow only in washes where they receive water and their seeds can be scarified, and Pickleweed grows best in alkaline soils. The presence of certain indicator species tells about the general ecological characteristics of an area, and plant communities are often named for the indicator species.

THE CHAPARRAL is the first vegetational community encountered after crossing the border. This Chaparral is a continuation of this vegetational community in the southern portion of California. The name of this plant community is derived from the Spanish word Chaparral which means evergreen Oak. This community is composed of dwarfed woody trees and shrubs covering the hills and lower mountain slopes of the Pacific coast of Southern California and the western slopes of Baja California from the International Border to Valle Santo Tomas.

There are two distinct kinds of Chaparral in the Californian phytogeographic region of northern Baja. Both types of Chaparral are important in flood control and in the prevention of soil erosion of the watersheds.

The **COASTAL SAGE SCRUB** is located at lower elevations along the Pacific coast. The flora of this type of chaparral is short (.3 to 2 m. tall) and less dense than the true chaparral. The dominant indicator species of the Coastal Sage Scrub are: California Sagebrush, Purple Sage, Black Sage, Lemonade Sumac (the only tall plant), Yarrow, and several species of wild Buckwheat.

Located inland from the Pacific coast the **TRUE CHAPARRAL** is found at slightly higher elevations on the hills and western slopes of the two mountain ranges of northern Baja (below 1,000 meters) southward to Valle Santo Tomas. True chaparral is made of stiff-branched, small-leafed, densely growing, woody trees and shrubs. The trees and shrubs are somewhat taller than those of the Coastal Sage Scrub Chaparral (1 to 3.5 meters). Plants of the True Chaparral are adapted to long hot summers. The True Chaparral indicator species are: Chamise, Toyon, Buckthorns, Scrub Oak, Mountain Mahogany, Yucca, Ceanothus, and Manzanita.

Several common Chaparral species were used by Indians in both of the Californias. An acidic tea was made by soaking the fruit and seeds of the Lemonade Sumac (*Rhus*) in water. Leaves of the Wild Buckwheat Brush (*Eriogonum*) was used variously as a decoction for headaches and stomach pains A tea from the flowers of Buckwheat was used as an eye wash, and the stems were boiled for a tea to treat bladder problems.

12.5 The hills to the left are the fossiliferous Pliocene-Pleistocene sandstones and conglomerates of the San Diego Formation.

13 The first exposures of the volcanic and sedimentary rocks of the Middle Miocene Rosarito Beach Formation are located near the southern end of the Playas de Tijuana. They were first defined by Minch (1967) as a series of basalts (a red-brown to black, fine-grained, iron rich igneous rock), tuffs (a fine-grained fragmental rock of volcanic ash), and tuffaceous sedimentary rocks (fluvial deposits of tuffs) of Middle Miocene age (14.5 m.y.) which are exposed between Tijuana and Rosarito.

The corn, beans, and other vegetables growing on the hillsides along the left side of the highway are being watered with sewage effluent which provides water and fertilizer for these crops.

ROSARITO BEACH FORMATION IN SEA CLIFFS

14 The hills to the left are largely composed of the basalts and tuffs of the Rosarito Beach Formation of Miocene age with a thin veneer of sandstones and conglomerates of the San Diego Fm. of Pliocene-Pleistocene age. The San Diego Formation was deposited in a near-shore environment. The Pliocene shoreline above the highway is a few hundred feet further inland and a few hundred feet higher than the present day shoreline. Since the Pliocene the shoreline has moved westward as the area was uplifted.

14.8 At the La Jolla turnoff the strata (rock layers) exposed in the roadcut are composed of volcanic tuffs overlain by basalts. These tuffs and basalts are part of the Costa Azul Member of the Rosarito Beach Formation. A prominent bake zone of reddish tuff is exposed in the outcrop. It formed in the Middle Miocene as molten basalts flowed over the white tuffaceous material and baked them to extremely hot, 2000° F, temperatures.

ROSARITO BEACH FM. BASALTS OVER TUFFS

The La Jolla turnoff provides an opportunity to observe the lithologies and relations between two members of the Middle Miocene Rosarito Beach Formation and the Upper Pliocene San Diego Formation.

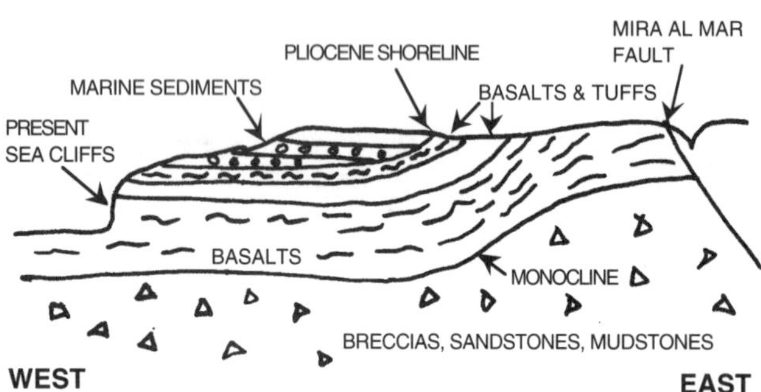

FRANCISCAN-TYPE DETRITUS: The base of the section is located about 0.5 kilometers up the arroyo to the east around the southeast side of the ridge. Exposed there is the lowest, Mira al Mar Member, of the Rosarito Beach Formation. This member is exposed in the core of a monocline and represents mudflows composed of a light-gray, medium to coarse-grained, arkosic, sandy matrix breccia. These breccias contain fragments of Franciscan-type detritus such as glaucophane schist and other schists, serpentine, sausuritized gabbro, bedded chert (a dense, hard, siliceous rock), quartzites, and minor amounts of acidic, volcanic and plutonic rocks (Minch, 1967). In another canyon an oolitic limestone (spheroidal or ellipsoidal particles formed by chemical precipitation in shallow, wave-agitated water), and tuffaceous sandstone are exposed in the section indicating a relatively quiet deposition between the lobes of this submarine mudflow fan complex.

This member has been interpreted to represent a series of mudflow deposits originating from a western landmass which were deposited on a narrow shelf against and onto the slope of the landmass area to the east. Between mudflows ocean currents and wave action reworked the mudflows which produced sandy matrix breccias next to the lobes, and sandstones and shales more distant to the lobes. This area, between the offshore volcanic highlands and the Baja California mainland, was periodically swept by strong currents. This unit is exposed along this arroyo and in canyons along the Los Buenos Fault in the Tijuana-Rosarito coastal area (Minch, *et al.*, 1984). This member is correlated with the Los Indios Member exposed to the south in the La Mision area.

PLIOCENE SHORELINE

A short walk along the dirt road and to the left leads to a fossil locality in the near shore Pliocene San Diego Formation which overlies the basalts and tuffs of the Coastal Azul Member at this site.

The fossils occurring at this site (and in another notable locality in the hills east of Km 15.2) are found in poorly-sorted yellow-brown conglomeratic sandstone beds resting directly on the weathered surface of the eastward-sloping Miocene basalts.

Do not collect Baja's fossil materials since they are all needed to illuminate Baja's past. It is against Mexican federal laws to collect any of the peninsula's resources!

15.2 A richly fossiliferous San Diego Formation locality on the north side of this canyon just east of the highway yielded an invertebrate fauna of 36 species from lensing sandstone beds about 6 feet above the contact with Miocene volcanic rocks.

A MARINE PLIOCENE SHORELINE AND FOSSILS: The rich fossil fauna of the two localities (Kms 14.8 and 15.2) is dominated by shells and shell fragments of the extinct molluscs: *Pecten healeyi*, *Acanthina emersoni*, *Anadara trilineata* and *Chalmys parmeleei*. These extinct fossil species are characteristic of Pliocene strata found on both sides of the International Border.

The environmental living conditions of both the extinct and extant (living) species of mollusks in this region was one in which the ocean temperatures were cooler than those presently encountered in the San Diego-Tijuana region. Studies comparing past and present environments are extremely important to geologists and biologists, as they seek to reconstruct the past history of the earth and predict its future.

The fossil fauna found at the two localities can be divided into two cool water components. The first is a **littoral** and **inner sublittoral epifauna** which lives on an exposed coastal rocky substrate. This component is represented by the indicator genera: *Acanthina*, *Calliostoma*, *Balanophillia*, *Penitella*, *Tegula*, and *Thais*.

The second component consists primarily of the following **sublittoral** indicator genera: *Acila*, *Dentallium*, *Dosinia*, *Laevicardium*, *Nuculana*, *Panope*, *Protothaca*, *Siliqua*, *Spisula*, *Terebra*, and *Tresus*. The environmental requirements of these genera suggest a sublittoral, semi-protected silt or sand substrate environment. The mollusc *Calyptraea mammilaris*, a warm water organism, indicates that a warm infaunal element also existed as a subcomponent of this cool water environmental element.

Also present at this locality is a substantial vertebrate fauna which includes the large teeth of *Carcharadon megalodon*, the huge extinct cousin to the modern Great White Shark (*Carcharadon carcharias*). The presence of this species at this locality is the first published record (Ashby and Minch, 1984) from the late Pliocene. *Carcharadon sulcidens*, another Great White Shark, is also present, as it is in many Pliocene sedimentary rocks throughout Baja and California. This species has been synonymized with the modern Great White Shark.

Other vertebrates present at this locality include whale and dolphin remains, a Mako shark (*Isurus planus*), a bay shark (*Carcharias sp.*), and a bay ray (*Myliobatis sp.*). *See* Ashby and Minch (1984) and Rowland (1972) for more detailed discussions of the paleontology at this locality.

15.5 The upper part of the roadcut on the left exposes a series of prominent cross-bedded Pliocene sandstones. These overlie Miocene basalts of the Rosarito Beach Formation. A whole rock K/Ar radiometric date on this basalt gave an experimental age of 14.3 $\pm$ 1.2 million years. This date is probably close to the true age although the moderate degree of alteration of the sample poses some uncertainty. In general, it seems reasonably close to the widespread 15 m.y. ages for basaltic vulcanism on the Continental Borderland and these rocks are probably representatives of a volcanic province which was active in mid-Tertiary time and extended as far westward as the continental slope (Patton Escarpment), north to the Transverse Ranges, and eastward to the Gulf Of California (Hawkins, 1970).

BEDDING AND CROSS-BEDDING: Stratification is a sedimentary rock structure (commonly referred to as layering or bedding) and is the most distinctive feature of sedimentary rocks. In general, each bedding plane marks the termination of one period of deposition and the beginning of another. The layers (strata) seen in sedimentary rock are formed when a sediment is laid down. They are later revealed by exposure due to weathering and erosion. Bedding is usually horizontal, but cross-bedding occurs at some angle when wind or water deposits a sediment on a slope. The cross-bedding at this site resulted from erosion of the beds by waves or shifts in the ocean currents along a rugged coastline.

Subsequent beds were deposited at an angle to the eroded beds.

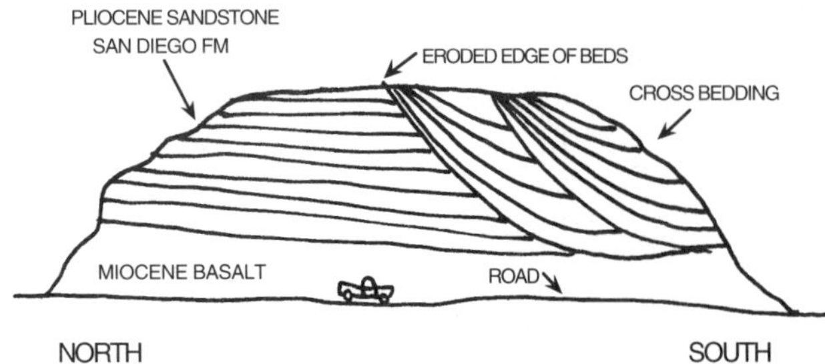

16.1 From the Punta Bandera turnoff the highway follows a narrow strip of the Late Pleistocene Terrace all the way to Rosarito. From this point good exposures of the Miocene fluvial sedimentary rocks are visible on the middle rock of the north island and the south island of Islas Coronados.

A short walk north along the beach at this turnoff reaches excellent exposures of airfall lapilli tuffs and basalts in the Costa Azul member of the Rosarito Beach Formation.

19.0 The highway crosses Canon San Antonio de Los Buenos. In 1997 the Tijuana sewer utilized this canyon as an access to the ocean. Three Kms. up this canyon the Mira al Mar Member of the Rosarito Beach Formation contains Franciscan detritus (*See* 1:14.8) and Miocene fossils.

Fossils from this well-developed Pleistocene terrace consist of shallow water marine invertebrates mixed with aboriginal (early man) and land mammal remains.

In this region the Nestor Terrace was very heavily occupied by several Indian cultures. Evidence of prior habitation is found in numerous areas along the terrace in the form of "kitchen or trash middens".

INDIAN KITCHEN OR TRASH MIDDENS: On the north side of Canon San Antonio de Los Buenos an extensive Indian trash midden is exposed in the cliff just above the highway. The upper dark soil layer represents remains of aboriginal occupation of the terrace. For many years native Indians lived along the Pacific coastline. They did not have permanent villages, but lived together in small family camps which they moved when the area was "fished out" or became infested with fleas (pulgas). In this way the camps moved up and down the coast and over the years an almost continuous layer of midden material was deposited. Some anthropologists believe it is possible that Indians may have occupied this particular area as early as four thousand years ago. Excavations at midden sites like these are very valuable to archaeologists and anthropologists; they provide many man-made artifacts, burials, and plant and animal organic remains which help to reconstruct their life-style.

Do not disturb the midden or remove <u>anything</u> from Baja. It is **ILLEGAL**! (*See* 3:171 for a discussion of the prehistory of early man in Baja.)

22.0 This Late Pleistocene Terrace near San Antonio Shores has also experienced severe sea cliff erosion during the last three decades.

Excavations on the Nestor terrace at San Antonio shores have yielded the remains of a possible new species of mastodon (*Stegamastodon sp.*), a relatively recent relative of the modern horse (*Equus caballus*), and other mammal bone fossils. These fossils represent a land fauna which occupied this low coastal terrace following its late Pleistocene emergence.

22.5 The roadcuts for the next 11 Kms. expose basalts and tuffs of the Rosarito Beach Formation. The reddish tuffs near the top of this section were baked by the Miocene basalts (*See* 1:14.8).

23 The gently rolling hills to the left of the highway are developed on the basalts of the Rosarito Beach Formation which weathers to a very rich fertile clay soil. The lighter colored sedimentary rocks in this area are sandier tuffaceous beds which do not weather into good soil on the slopes in this area. To the south near La Mision almost the reverse occurs. The Los Indios member of the Rosarito Beach Formation, which is largely clayey tuffaceous sedimentary rocks, is the one most farmed. There the clayey tuffs form good soil on the flat areas and the basalts form very rocky areas.

25 As the highway makes a bend, observe the rugged shoreline. The surf along this stretch of the coast is often quite spectacular due to the rougher topography which makes the waves crash on the shore. This often produces a loud "boom"!

27 The Rosarito Power Plant to the right burns fossil fuels to supply much of the electricity for northwestern Baja California. Tankers delivering fuel to the plant are often seen moored offshore.

BASALT CAPPED MESA REDONDA

To the east is the prominent steep-sided flat-topped Mesa Redonda. The prominent peak to the right of Mesa Redonda is called Cerro Colonel. Both hills are capped by Miocene basalts and are underlain by Eocene and Upper Cretaceous strata. Mesa Redonda was formed as the sedimentary rocks were eroded by streams which dissected the landscape. This left the flat-topped mesas standing above more erodible areas.

27.3 The highway descends a fault scarp as it crosses the active Agua Caliente Fault which offsets the Late Pleistocene terrace in this area. This fault was given this name because it runs through the Caliente Race Track and is responsible for the Agua Caliente Hot Springs at the race track. The Agua Caliente Hot Springs is one of a number of hot springs located along faults in Baja California (*See* 13:176). This fault also runs through the area of the Rosarito Power Plant and tank farm.

THE CALIFORNIAN REGION AVIFAUNA: The bird species of the Pacific coast of Northern Baja are the same as those found along the Pacific shoreline and in the Chaparral and Oak-Woodland plant communities of the Californian Phytogeographic Region of Southern California.

Along the Highway where a species of bird has been frequently seen, a brief natural history discussion will be presented for that bird along with a line sketch of it, a discussion of distinctive identifying characteristics, and the Mexican name.

BIRD NAME	LIKELY LOCATION
Birds found throughout Baja in suitable habitats:	
American Kestrel	wires and fence posts
American White Pelican	Gliding along the shore
Anna's Hummingbird	Feeding on red or yellow tubular flowers
Ash-throated Flycatcher	Desert, chaparral, woodlands
Black-chinned Hummingbird	Feeding on red or yellow tubular flowers
Black-tailed Gnatcatcher	Flying and flitting through low chaparral brush
Blue-gray Gnatcatcher	Flying and flitting through low chaparral brush
Cactus Wren	On cacti
California Brown Pelican	Gliding along the shore
California Quail	On the ground
Common Raven	Flying and perching on trees, fences, telephone poles
Greater Roadrunner	Crossing the highway
Hooded Oriole	Common around palms
Horned Lark	Dirt fields, gravel ridges, and shorelines
House Finch	Abundant in all areas
Ladder-backed Woodpecker	Flitting in the air
Northern Mockingbird	Thickets, woodlands, towns
Red-Tailed Hawk	Tops of telephone poles and fence posts
Rock Wren	Scrublands, dry washes, and most arid areas
Turkey Vultures	Soaring in the skies or feeding on carrion on highway
Verdin	Mesquite and thorny shrubs
Western Meadowlark	Fence posts and fence wires
Birds of the California Region:	
Bell's Vireo	In moist woodlands, bottomlands, mesquite
Brown Towhee	In brushy hillsides, wooded canyons, and chaparral
California Thrasher	On the ground
Killdeer	On the edges of marshes, shorelines, and salt flats
Lawrence's Goldfinch	In the chaparral and dry, grassy areas
Loggerhead Shrike	Wires and fence posts
Nuttal's Woodpecker	Chaparral and oak woodlands
Plain Titmouse	oaks, junipers, pinyons, pines
Red Shafted Flicker	Chaparral and oak woodlands
Scrub Jay	Chaparral and oak-woodlands
Starling	Flocking on telephone wires, trees, grain fields
Wrentit	Chaparral, coniferous woodlands

27.7 Excavations in a road metal quarry to the north of the highway have yielded a diverse and abundant Late Pleistocene assemblage of marine invertebrates. White sandstones high on the cut at the northeastern end of the quarry near the crest of the hill into which the quarry is excavated yielded 55 species of invertebrates at two localities. No southern faunal forms have been recognized. A conglomerate on the floor of the southeast face yielded only 7 species and includes the extinct mollusc *Crepidula princeps*.

29.4 The Rosarito turnoff is a business loop through the town of Rosarito that eventually returns to the toll-road near Km 35.

29.8 A road from Tijuana joins the toll-highway. The highway continues south on the Late Pleistocene terrace cut on the Miocene volcanic sequence.

33.6 Just south of town is the long-standing Renee's Rosarito Beach Motel for which the Rosarito Beach Formation was named.

34.5 The road from Rosarito returns to the toll highway where the free road leaves the highway. The free road parallels the highway to La Mision. The free road provides access to local beaches. This road log (guide) follows the toll road

35.5 Caseta de Cobro

36 South of the toll station, the highway follows another Pleistocene terrace developed on the Rosarito Beach and Rosario Formations which is locally interrupted by landslides where weak sedimentary rocks underlie the capping basalts.

39 There is a prominent hill to the left of the highway with what looks like vertical columns or posts. They are the result of a geological phenomenon known as columnar jointing. This hill is the neck (volcanic plug) of a volcano younger in age than the Rosarito Beach Formation. (It was not a source for the volcanic rocks of the Rosarito Beach Formation). It is one of a number of volcanic plugs that dot the International Border area from Tijuana to the east for tens of kilometers.

VOLCANIC PLUGS are masses of rock which seal the vents and conduits of volcanoes. These vents become exposed as the more erodible surrounding rock of the original cone is removed.

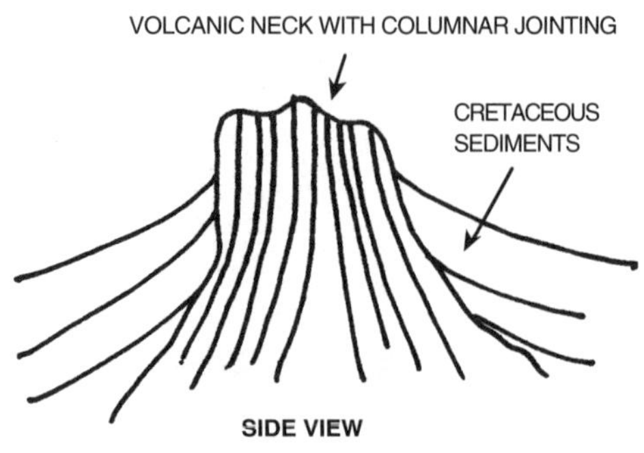

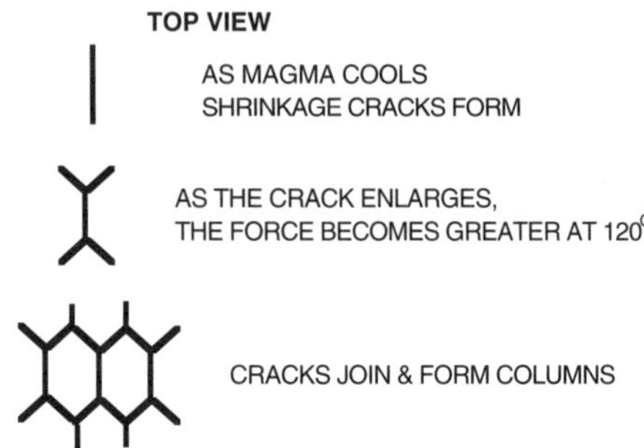

42 The outcrops in the hills are the marine sedimentary beds of the Cretaceous Rosario Formation.

This Pleistocene terrace was once more extensively developed. As a result of erosion of the softer sedimentary rocks, the terrace has been reduced to remnants with intervening areas weathered into rounded hills and gullies.

Late Pleistocene marine terrace deposits rest on Miocene volcanic rocks here and elsewhere along the local coastline. On the west side of Punta Descanso just north of the point about 2 meters of fossiliferous material is exposed adjacent to the cut terrace surface along about 40 meters of sea cliff. The lower one meter of that deposit is an unconsolidated pebble conglomerate in a sand matrix. The upper one meter is unconsolidated rubble with a matrix of shells. These two rock layers have yielded about 150 invertebrate species. They are chiefly molluscs, including *Chione picta* (living from Bahia Magdalena southward) and *Velutina laevigata* (living north from Cayucos, California).

44.5 An Indian midden is exposed in the roadcuts on both sides of the road for the next 0.5 km. If left undisturbed these middens will provide valuable information about the history of man in Baja as future archaeological studies are conducted.

45 The basalts of the Rosarito Beach Formation overlay the Rosario Fm. and are well exposed in some of the cliffs east of the highway.

The poles and wires in this region are usually good places to look for American Kestrels.

American Kestrel or Cernicalo Chítero is the smallest and most vocal among Baja's birds of prey and is the most commonly seen falcon of the open country in Baja. This is also the only falcon with a rusty back and tail. It hunts from poles, wires, or trees and is seen hovering in the air before it stoops to capture some ground-dwelling insects, reptiles, or rodents.

46 The brick-making operations left of the highway utilize the clays of the Rosario Formation. The bricks are formed in wooden molds and sun dried. They are stacked in the shape of a loosely filled hollow "room" which is plastered with mud, filled with firewood, and lit. The fire hardens the bricks.

46.5 For the next four kilometers there are good views to the east of Cerro Colonel. The small basalt mass offshore is known as Moro (snout) Rock because it resembles a nose.

48 For the next kilometer there are good exposures of the Rosario Formation in roadcuts along both sides of the highway and in the sea cliffs.

49 This is the turnoff to the beach front communities of Cantiles and Puerto Nuevo.

52.2 The highway crosses Valle El Moro. The bridge over the highway provides an unpaved access to the valley from the old road. A rock quarry located in this valley is easily reached by a high-clearance vehicle.

ROSARIO FORMATION IN SEA CLIFFS

Watch for California Quail (*Callipepla californicus*) on the drive into this valley.

California Quail or Codorniz Californiana: They are members of the pheasant subfamily and are small plump bodied intricately colored birds which live in the coastal and foothill chaparral, live-oak canyons, deserts, and oasis throughout Baja. They often travel and feed in large coveys. Their most identifiable field characteristics are their bobbing black top-knot, feathers, and scaled chests. Depending on the location of the accent the voice of these shy skittish birds seems to say "*come* right here" or "where *are* you?".

The quarry located in Valle El Moro is a **buttress unconformity** of the Rosario Formation against a metavolcanic hill of the Alisitos Formation. It was formed as the mudstones of the Rosario Formation were deposited against a submarine hill composed of the resistant metavolcanic rocks of the Alisitos Formation.

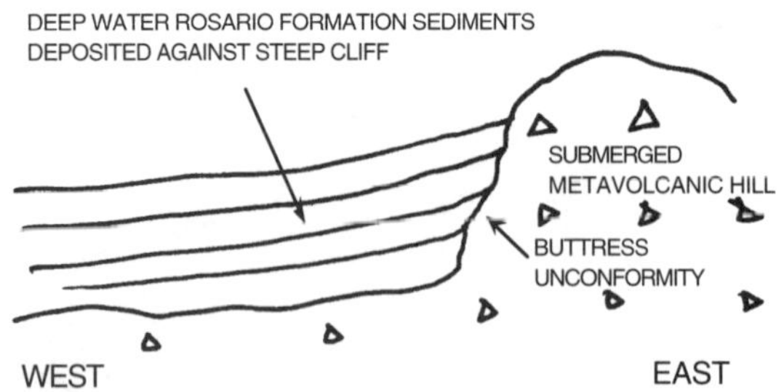

BUTTRESS UNCONFORMITY AT EL MORRO

54.5 The Medanos coastal dune field formed as a result of ocean sands blown on shore by strong winds coming from the open sea. The dunes are stabilized (they don't migrate) by the roots of the vegetation which grows on them. Sometimes the dunes can be "blown out" by wind or exceptionally high waves. If this occurs, they start to move and are no longer stabilized. When the dunes are "blown out" a secondary type of plant succession known as "old dune succession" occurs as vegetation returns to stabilize the dunes.

55.5 For the next 40 Kms. between here and San Miguel the highway will skirt the high mesas of the La Mision Member of the Rosarito Beach Formation, passing between them and the Ocean.

BASALT CLIFFS IN DESCANSO VALLEY

56 The small white church on the hill north of Arroyo Descanso is on the former site of Mision Descanso. It was established in 1814 as the northernmost Dominican mission. It was the next to the last mission established in Baja California.

58.6 Cuenca Lechera turnoff. This name probably refers to the nearby dairy. It provides access to Medio Camino and the free road.

59.4 Medio Camino (Half Way House) is located halfway between Tijuana and Ensenada.

The Half Way House is built on tuffs and basalts of the Rosarito Beach Formation. The mesas visible to the east of the highway are capped by the Rosarito Beach Formation overlying the Cretaceous Rosario Formation. The gently-dipping tuffs of the Punta Mesquite Member of the Rosarito Beach Formation outcrop in the sea cliffs north and south of this point. The lithic tuffaceous sandstones beneath the Half Way House are slightly different in composition from the tuffs to the north. They are coarser and contain interbeds of siltstones and shales. They have been highly faulted. A walk along the beach to the north reveals exposures of this tuff.

Examples of sedimentary structures that can be examined include: channel cross-bedding, hummocky cross-stratification, channel cross-stratification, load structures, rip-up clasts, and bioturbation (*Omphiomorpha*) of various kinds.

This area is interpreted as proximal to the offshore volcanic highlands during the Miocene.

Deposition of these sedimentary rocks was above storm wave base (Hummocky cross-stratification) and well-oxygenated (bioturbation). The sedimentary structures all exhibit an eastward transport direction. This area contains the oldest exposures of the Rosarito Beach Formation. Preliminary findings indicate that these tuff exposures represent a portion of the western margin of the basin during the Middle Miocene.

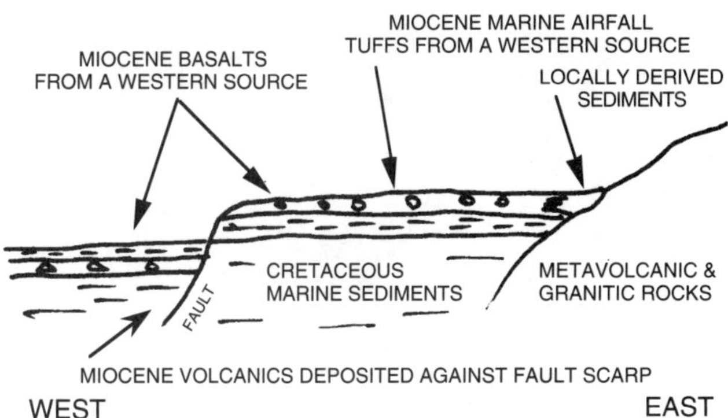

61 In late spring (May-June) the gray-white "dots" of vegetation growing on the hillsides to the left of the highway are called Siempreviva or Live-Forevers (*Dudleya pulverulenta*).

LIVE-FOREVERS are perennial, herbaceous succulents which commonly grow in the poorer rocky soils in the Coastal Sage Scrub and foot-hill chaparral communities of the beaches and coastal bluffs, south to El Rosario.

62 To the north is a panoramic view of the coastline with Punta Descanso in the distance. The tuffs of the Rosarito Beach Formation are exposed in the near sea cliff. On a clear day the southern island of Islas Coronado is visible offshore.

62.3 Slump blocks are common along the highway for the next 3 Kilometers. The closed depressions are unaltered by drainage. Many of the slumps project into the sea like brown gnarled fingers.

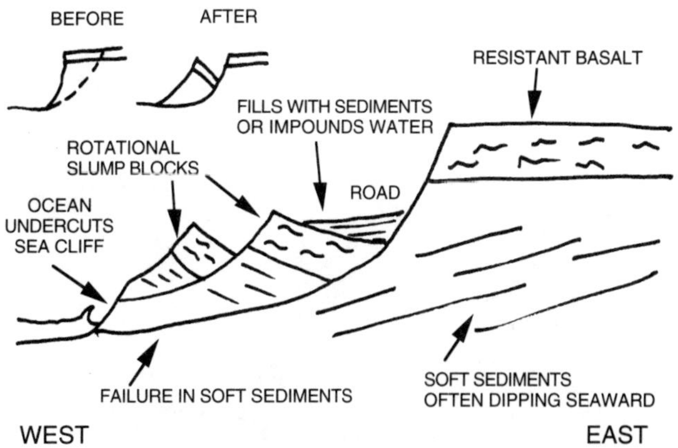

Marine erosion and slumping has formed high cliffs in the thick sequence of soft Cretaceous sedimentary rocks. This has resulted in the undermining of the resistant Miocene basalt caps. Failures along the coastline of this region (slumps and landslides) are influenced by the thickness of the resistant volcanic units, the lithology of the underlying sedimentary rocks, the north-south jointing and faulting, the angle and attitude of the bedding planes, and groundwater. The dominant cause of the failures is the inherent weakness of the underlying mudstones and over steepening caused by the marine erosion. Most of the slides are rotational slumps; however, some sliding may be due to block glide where the slope of the beds is seaward. The toe of the slide block is commonly uplifted offshore. During calm seas muddy water may be found at the toe of a recent slide .

64 To the right is Plaza del Mar with a replica of a Mayan Pyramid. It is built in the middle of a massive landslide bowl. This area exhibits a low hilly (hummocky) topography and indicates that many landslides have occurred here. At various times in the distant past the whole coastline has slid downward and westward. Locally, prominent terraces have developed on the La Mision and older members of the Rosarito Beach Formation which attests to the relative stability of some of these landslides. Many of them occurred in excess of 50,000 years ago.

Coastline slump block failures: South of this area the toll highway crosses Quaternary to

Recent aged landslides for approximately 40 kilometers of rugged coastline ranging up to 500 feet above sea level. Most of the seacliffs are underlain by a thick sequence of flat to gently southwest-dipping upper Cretaceous mudstones, sandstones and conglomerates of the Rosario Formation. These softer and more easily erodible sedimentary layers are by resistant Miocene basalts of the Rosarito Beach Formation.

SLUMP FAILURE ONTO HIGHWAY

66 The La Fonda turnoff is the last turnoff for traffic between the toll road and the free road. The toll road is the shorter route to Ensenada. The La Fonda Resort is a popular tourist area situated on a basalt seacliff on the very narrow Pleistocene coastal terrace. Demetri is one of my favorite hosts and his terrace dining is a must. Dolphins are often seen in the surf and Pelicans have flown by less than three meters above my head.

Look for the pelicans which are commonly seen skimming close to the waters just offshore.

The **American White Pelican or Pelícano Blanco** and the **California Brown Pelican or Pelícano Moreno** are the large aquatic fish-eating birds seen along both coasts and on many of the gulf islands in Baja where they nest in large colonies. They are commonly seen flying in long straight lines (brown pelican) or "V" shaped formations (white pelicans) only centimeters above the surface of the water. Feeding is accomplished by diving into the sea from as high as 50 meters (brown pelican) or by bill-scooping as they wade or swim in shallow coastal or lake waters (white pelicans).

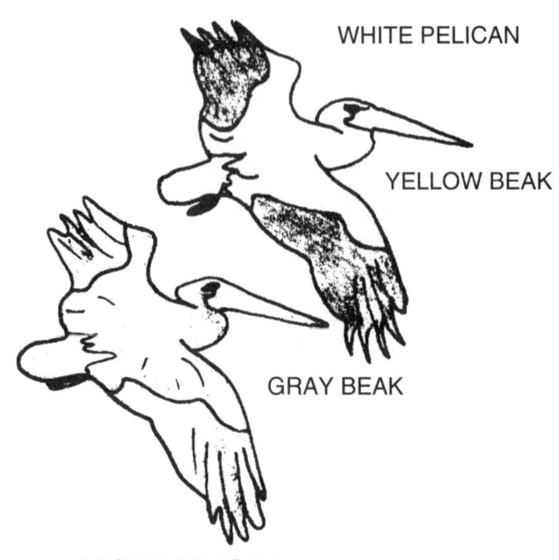

The slow deliberate flight of the brown pelican, low over the water with sudden plunges for fish, makes its identity unmistakable. The brown pelican is commonly on fishing piers and around docks. *A funny bird is the pelican, his beak can hold more than his belly can.*

67.5 A view of the older development of La Mision reveals a fine flat beach and a scenic rocky headland. At one time this was the only developed area along the highway between Tijuana and Ensenada. The roadcuts in this area are the basalts of the La Mision Member of the Rosarito Beach Formation.

69.2 The highway crosses the estuary of the Guadalupe River which originates in the high Sierra Juarez near Laguna Hanson. Deep wells in this valley provide part of the water supply for the city of Tijuana. The surge tower of the pipeline can be seen on the hill north of the estuary.

Coastal Wetlands are **Estuarine Environments** which exist where the ocean tides meet the outflow of a river current. The water of this environment may be alternately fresh or brackish. The marsh and aquatic plants are of great value to shorebirds, to some small mammals, and to songbirds such as the Red-winged Blackbird, Sparrow, and Marsh Wren. The marsh is rich in plant and animal species.

GUADALUPE RIVER AT LA MISION WITH SALT MARSH AND BASALT MESAS

The characteristic plants of this environment are members of the rush family. Spiny Rushes (*Juncus acutus*) are tufted grasslike herbs commonly found growing in moist places like the Guadalupe Estuarine Marsh. The durable stems of this herb were used by early man in the construction of baskets. Other plants of this marsh are Yerba Mansa, Salt Grass, Alkali Heath, Heliotrope, Sea Lavender, Salt Cedar, Glasswort, Pickleweed, Sea Purslane, and Sea Blite.

The birds most often seen here are the Killdeer, Red-winged Blackbird, Egret, and Great Blue Heron.

The **Killdeer or Tildío**, is widespread throughout the entire peninsula. Because of its plaintive lonely cry, *kee-kil-de-dee*, early Spaniards named this bird El Perdido, which means "the lonely one." It is often seen and/or heard in agricultural fields, short grassy areas, and along the borders of salt marshes and the shorelines of both of Baja's coasts. The distinctive identifying characteristics of the Killdeer are its double black-banded white neck and orange upper tail and lower back. It performs a "broken wing routine" when nests or young are threatened. Their open depression-like nests are usually built on gravelly soils.

The **Red-Winged Blackbird or Tordo Charretero** is a resident of fresh water marshes, fields, and moist grasslands in northern Baja. The males are easily recognizable by their red shoulder epaulettes. The epaulettes are used to defend their territory from other male Red-Wings. This abundant, aggressive species is often found in immense flocks in winter.

The **Great Blue Heron or Garza Azul** is often seen in the marshes or on the beaches, or even in the dry fields of Baja. The tall, lean solitary figure of the Great Blue Heron can be seen standing motionless in a pool of water or advancing slowly one step at a time, lifting each foot stealthily from the shallows without a ripple. Herons may stand as still as a statue for over half an hour while waiting for prey. With a lightning-quick forward lunge of their long neck and bill the heron captures its prey. They prefer fish but will also eat birds, small mammals, insects, snakes, frogs, and crustaceans.

Whether it is on land or in the air, a Great Blue is easily recognized by its yellowish bill, ornate plumes on its head, neck, and back, the long snakelike neck held in an "S" shaped curve, the slow wing strokes, long legs, and nearly 2 meter wing span. In Baja look for the solitary Great Blue Heron on piers, docks, sandbars, or in estuaries, coves, marshes, and the riparian woodland.

The **Great (Common) Egret or Garzón Blanco** resides throughout the lowlands of Baja near fresh water streams, salt marshes, ponds, mudflats and estuarine environments. This is one of the larger wading birds in Baja, larger than any other heron except the Great Blue Heron (*See* below). The Great Egret has white plumage, a yellow bill, and shiny black legs and feet. Like the Great Blue Heron, the Great Egret is a slow moving patient hunter of shallow water fish.

69.5 An excellent turnoff alongside the highway provides easy access to a wide and flat beach. Many beaches form where there is an abundant supply of sand, such as the Guadalupe River. Waves distribute them along the coastline.

70 The basalt forming the roadcuts here and on which the coastal homes in the area are built may represent a massive slump block or a fault block. An alternate theory is that this basalt was deposited against the Cretaceous seacliff and is older than the basalt which forms the mesa top. The overlying tuffs were stripped from much of the coastal area which left a flat surface along the coast for several kilometers. The Late Pleistocene Terrace (50,000 years old) has developed on top of this and shows the antiquity and stability of some of the coastline between Tijuana and Ensenada. Further south the coastline is highly unstable and is actively moving downward and seaward. This is demonstrated by the numerous repairs that are continually being made in the undulating "bumpy" highway.

72 On the right is Baja Ensenada RV Trailer Park. Note super-tidal area. It is likely that during some future particularly violent Pacific storm, this park may be inundated by high tides.

The Nopales cacti growing along the highway on the left were planted for use as food. (*See* 14:10).

73 The tidal flat (salt marsh) of La Salina just south of the turnoff is also a super-tidal area. Frequently the mouths of many of Baja's coastal river valleys are semi-sealed by a sand bar (stretches of beach, or dunes that separate the open sea front from the tidal flats to the rear.) This provides for the tidal flat with occasional access to the sea.

At one time this tidal flat must have been very popular with the Indians, as fairly extensive Indian kitchen midden (trash) materials extend eastward from the highway for about a kilometer on both sides of the marsh. The highway turnoff cuts through and exposes a small portion of this vast midden. A very large (12 cm.) perfectly worked spear point of black metavolcanic material was discovered in the triangle of ground formed by the two roads which enter and exit the highway. (Refer to 3:17 for history of man in Baja.)

TIDAL (MARSH OR MUD) FLAT ENVIRONMENTS: In general Baja's tidal flats occur between mean high tide and mean low tide levels, are vegetated by unicellular and larger forms of algae, and are bordered on their inland edge by intertidal Pickleweed and other halophytes (salt loving plants). Intertidal Pickleweed begins its best growth at the average high tide line.

Tidal flats are the home or favored resting place for many living organisms adapted to life in this extremely harsh saline environment. Organisms that inhabit tidal flats face severe changes in daily environments. They are alternately immersed by high tides or exposed by low tides. Part of the tidal flat will always be underwater in sloughs and channels, and part will be above the high water mark. Because of these changing conditions, variations in soil salinity are to be expected.

In the flattest lowest portion of the mud flats the salt concentrations may exceed 6% which prevents the growth of all plants except marine algae. Near the shallows and edges of the flats, salinity both decreases and fluctuates with fresh water runoff and rain. Temperature ranges are broad which necessitates adaptations for surviving fluctuations of as much as 50° F in a single day's cycle.

Closer to the shore, away from the shallows and on the banks bordering the tidal flat, halophytic plant roots capture silt, mud, and sand. The halophytes which can be seen growing around La Salina marsh divided into two communities based on their relationship to the tidal flat. The **dune and strand line flora** which is predominated by halophytic Sand Verbena, Sand Bur, Beach Evening Primrose, Beach Fig, Door Brush, and Ice Plant; and the **salt-marsh flora** predominated by halophytic Yerba Mansa, Salt Grass, Alkali Heath, Heliotrope, Spiny Rush, Sea Lavender, Salt Cedar, Glasswort, Pickleweed, Sea Purslane, and Sea Blite.

Further inland as the land slopes up to the surrounding hills covered by true Coastal Sage Scrub, the more salt-tolerant plants are replaced by those which are less tolerant.

Although they are not very obvious, animals thrive in mud and tidal flats (salt marshes). Due to increased nutrient flows, tidal flats are very fertile and are more prosperous than many other natural communities. Tidal flats are busy places especially for birds, such as Pintails, Mallards, Coots, Gulls, Terns, curlews, Willets, Dowitchers, Yellow Legs, and Whimbrels open mollusc shells and search for worms and shrimp in the tidal flats.

NO-SEE-UMS: Diptera (flies) is the fourth largest insect order. Dipteran mouthparts, which display the widest range of any insect order, range from piercing-sucking to sponging-lapping. Nearly all adult flies feed on liquids especially plant juices and blood. Some of the bloodsuckers are important human and animal pests. Flies are divided into three suborders, but only one (sand flies) concerns the Baja traveler. Sand flies (*Phlebotomus*) would hardly be noticed except for a painful bite the females inflict in order to suck blood. Sand fly adults occur near water while their larvae live in moist soil.

76 The highway passes through a roadcut in the tuffs and tuffaceous sandstones of the Rosarito Beach Formation. The tuffaceous strata of the Rosarito Beach Formation has yielded a Middle Miocene marine fossil fauna which includes: *Chione temblorensis*, *Anadara topangansis*, *Turritella ocoyana*, and numerous marine vertebrates. These strata have also yielded a camelid which has been tentatively identified as *Oxydactylus* cf. *longpipes*. This combination of marine and nonmarine species in the same locality suggests a marine-nonmarine tie-in between the Middle Miocene Hemingfordian North American land mammal age and the West Coast Temblor marine molluscan stage (Minch *et al.*, 1970).

77.8 The coastal resort of Baja Mar at Jatay offers a golf course, villas for rent, and a restaurant. The golf course has been described by many players as being "very challenging" because the fairways are "islands" in a "sea" of Cactus, Agave, and Coastal Sage Scrub. It has been said that errant players are just as likely to find their ball impaled on a cactus as resting on the fairway.

78 Just south of Jatay the hills to the left of the highway seem to be quite straight (linear in relation to the coast) and have a concave upward slope. This concave upward slope is the headwall of a very large rotational slump block. The land that this portion of the highway is on has slumped down and rotated. All along the highway in this area there are numerous concave areas which are indicative of the rotation of numerous such smaller slump blocks.

84 The turnoff to El Mirador is on a dangerous curve. Exit carefully! The edge of a 300 meter escarpment at El Mirador provides an excellent view of several coastal features which are visible in this area and include a view of Islas de Todos Santos.

The panoramic view of the coast to the south shows the mudstones, sandstones, and conglomerates of the upper Cretaceous Rosario Formation overlain by the basalts of the Rosarito Beach Formation. Numerous slumps are evident along the coast. Some have dropped capping volcanic rocks to sea level where they now form resistant shore promontories. The metavolcanic spine of Punta Banda on the south side of Bahia Ensenada is often visible in the distance. It is a fault block bounded on the northeast and southwest sides by two branches of the Agua Blanca Fault (*See* 2:10). Islas de Todos Santos are the offshore Islands.

ISLAS DE TODOS SANTOS: The two islands of Islas de Todos Santos are composed of middle Cretaceous metavolcanic rocks of the Alisitos Formation overlain by basalts (La Mision Member) and sedimentary rocks (Los Indios Member) of the Rosarito Beach Formation. Also present on the islands is a well-developed, Late Pleistocene marine terrace. The waves breaking on the north end are reported to be among the highest in the world. Legend says that the Islas de Todos Santos were the inspiration for Robert Lewis Stevenson's "Treasure Island".

IRONWOOD AND INDIANS: Ironwood (*Olneya tesota*) is found growing on bajadas and in Wash Woodlands below 600 meters throughout the entire peninsula and the west coast of mainland Mexico. Due to its density this wood is very heavy and extremely durable. The Seri, at Bahia Keno in Sonora, have carved this wood into spear shafts, arrow points, and agricultural implements. Since the 1950's the Seri (originally inhabitants of Isla Tiburon in the Gulf) and enterprising Mexicans have carved beautiful enduring dark-red wood sculptures in a variety of animal shapes sometimes sold at this turnoff.

85 If you missed the turnoff to El Mirador at Km 84, this turnout has a view similar to that at El Mirador. This spot provides a good place to look at the flat terraces below the highway. Each terrace is a rotational slump block. The viewpoint produced by the rotational movement of a slump block is directly above one of these terraces.

COASTAL VIEW WITH TERRACES ON SLUMP BLOCKS BELOW HIGHWAY

SLIPPING AND SLIDING: Because much of the coast is slumping into the Pacific along this stretch of the highway, the pavement from here to San Miguel undulates considerably. Between Salsipuedes and San Miguel active sliding caused by marine erosion requires continuous highway maintenance to compensate for the seaward movements of large slide blocks. Portions of the highway have been down-dropped several meters. New pavement marks the lateral edge of several slide blocks, and fresh scarps several meters high in natural ground mark the up-slope boundary of active sliding. Preventative and corrective road work in progress includes unloading (removing a portion) of the head of the slide, draining groundwater in and beneath the slide, and rechanneling surface drainage to prevent further infiltration of water into the slide.

87.5 Salsipuedes: In Spanish aptly means "Get out if you can". Early sailors sheltered in the cove below but found it difficult to climb the steep sea cliffs. A road to the right leads to a small rancho and tourist fish camp.

Watch for Brown Pelicans which soar along on sea cliff thermals between here and Ensenada. Another common bird is the mourning dove.

Mourning Doves or Paloma Triste: These are small slim birds with long sharply-tapered tails and black spots on the upper wing. They are commonly seen in northern Baja in gleaning seeds among the stubble of cultivated grain fields in the late summer or fall. They are frequently seen in cattle grazing ranges on dry uplands and in many of the villages and desert areas of Baja. Except for an occasional American Kestral, Doves are the largest birds that commonly perch on telephone wires where their long sharply-pointed tails make them easily identifiable. As you reach the southern part of the peninsula, watch for a close relative, the White-Winged Dove.

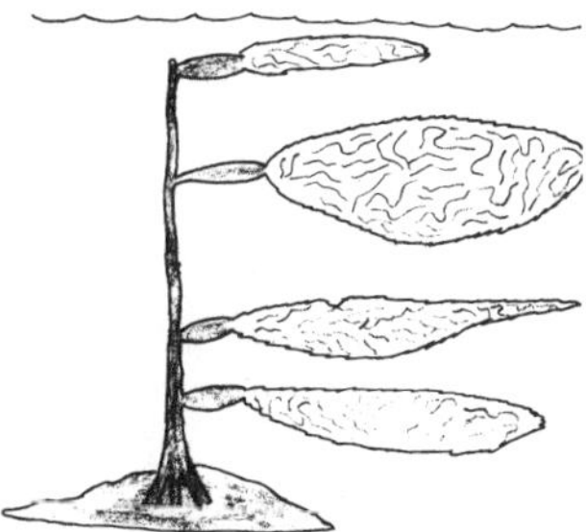

GIANT KELP is the common name for the Marine Algae *Macrocystis pyrifera*. It forms great kelp beds in the nearshore waters. As you look at the waters just off the beach, you may notice dark, caramel colored masses of vegetation floating near the seas surface. It is often referred to as the fastest growing plant, often growing .4 meters/ day. Giant Kelp beds provide an important habit of many marine organisms.

91 From here to Km 94 there are exposures of the massive sandstone and conglomerate lenses of the Rosario Formation. One of these submarine-fan systems is best seen east of the highway at Km 92. At this stop the sedimentary rocks record an upward transition from the slope (outer-mid fan) to the inner fan deposits. Here the section rests on a dark shale of the Rosario Formation (Middle Mudstone Member) that is overlain by a diamictite unit 12 meters thick. This is overlain by 7 meters of inversely graded imbricated conglomerate which represents slump and channelized debris flows. Above the conglomerates are 37 meters of submarine turbidite flows composed of sandstone and mudstone which fill a small submarine canyon. The overall thinning and fining upward nature of this sequence indicates an inner-fan channel of a submarine fan.

CRETACEOUS ROCKS NEAR ENSENADA: Sedimentary rocks deposited in this area during the Late Cretaceous are derived from the uplifted Peninsular Ranges Batholith. These sedimentary rocks developed as a clastic wedge, which is thicker to the west, and consists of a fluvial and alluvial facies to the east and a deep marine facies to the west. During Turonian time, fluvial and alluvial sediments were deposited as fans; these deposits are known as the Redondo Formation. The overlying sedimentary rocks comprise the Rosario Formation which are Campanian-Maastrichtian in age (Late Cretaceous), and the contact with the underlying Redondo Fm. is unconformable. In Campanian time sea level rose and deposited a transgressive sequence (beach to offshore deposits are represented). The shoreline was defined by steep bedrock cliffs with encrusting molluscan assemblages such as the rudistid-molluscan bivalve assemblage at Rincon de la Ballena (*See* 2:22:14, Punta Banda road). At the same time storm sedimentary deposits (hummocky cross-stratification) are recorded in the same area. The sea level rose with the deposition of a thick shale (mudstone) unit described by Kilmer (1963) in the type area until Late Campanian time. The sea level fell in Early Maastrichtian and formed huge deltaic sequences. This regression dumped large amounts of sediment on the steep narrow shelf and caused gravity-flow deposits which developed numerous submarine fan systems.

94.5 View point Turnoff - In the cove directly below a Mexican battleship tender was run aground at full speed during the night of Cinco de Mayo, 1967. The crew had mistaken the point at El Mirador for the one further south at Ensenada. It has been completely broken and washed away by the surf. Only one small beam was sticking above the surface of the water to mark the spot.

Remember the seven destroyers that were run ashore by the U.S. Navy at Point Arguello. Vessels of all nations have had trouble navigating the rugged Pacific coastline.

LANDSLIDE AREA OF PUNTA SAN MIGUEL CLIFF IS AT TOP OF SLIDE AREA

95.2 This is the site of a major landslide which disrupted travel along the highway for a number of months. During that time the old road was used as a bypass. Workers have stabilized the landslide by dewatering and repaved the highway. It has proven to be relatively stable since then.

96 The topography along both sides of the highway from here to Km. 98 is very rugged with many irregular small hills. Each hill represents an individual landslide area on the main point of Punta San Miguel. The entire coastline in this area is actively sliding into the sea due to the undermining of the point by the ocean.

98 This is the first view of the small fishing village of El Sauzal and the site of another massive highway failure which occurred in 1978. The slope failure on the hills to the left resulted in the loss of a number of residential homes which once overlooked the ocean. Look carefully to the left to see the remnants of some of these homes. These homes were built on landslide blocks. When the blocks failed (moved down toward the highway), the homes were destroyed. They were regrading the slide area and placing rock at the base of the slide in the summer of 1997. Several abandoned homes were still visible.

LANDSLIDE AT SAN MIGUEL

LANDSLIDE SURFACES offer attractive building sites to the unsuspecting developer because they are often flat with unobstructed views. The low density development on some of the stabilized blocks has not immediately created unstable ground. However, more development on presently stable landslides may create disastrous effects as shown by continuing problems in the San Miguel area. When structures are built on landslides, there is an increase in the ground water due to residential watering and waste disposal. When this additional water infiltrates into the landslide it acts to weaken, load, and lubricate the slide mass so that it becomes unstable and slides. The demand for beach-front property is increasing, and more of these stabilized blocks are becoming developed. More landslides, property loss, and possibly casualties are inevitable.

98.5 This is the last Caseta de Cobro.

LANDSLIDE AT SAN MIGUEL IN 1997 AFTER EXTENSIVE REGRADING

99.2 At the junction of the Tijuana Libre Highway and Highway 1D there is a view to the rear, of the remnants of the houses destroyed by the 1978 landslides and the neighboring houses on more landslides. San Miguel is to the right.

100 The highway between here and Ensenada follows the same Late Pleistocene Terrace which the road has been intermittently following since leaving Tijuana.

101 The hills to the left of the highway for the next several kms. are composed of upper Cretaceous marine sandstones which were deposited along an exposed rocky shoreline.

101.3 The turnoff to the right leads 114 Kilometers to Tecate on Mexico Highway 3 (*See* log 15).

101.6 The fish canning industry once supported the village of El Sauzal. A new harbor has been built here to shelter the fishing fleet. This was one of the enterprises of the late General Abelardo Rodriguez, a former Governor of Baja California and briefly the President of Mexico until the election of Lazaro Cardenas in 1933 (he replaced Ortiz Rubio in 1932 after Jefe Maximo Pluttarco Elias Calles forced Rubio to resign).

105 A quarry is located on the hillside east of the highway. The rocks which formed the Late Cretaceous coastline are exposed in the ravines immediately in back of the quarry and to the south.

106.5 The Universidad Autonoma de Baja California (UABC) with its Escuela Superior de Ciencas Marinas is located to the right of the highway on Punta Moro. This very prestigious marine biology school obtained legal rights to the property where the school is built after the students forced the governor of Baja to give it to them. UABC is one of the most prestigious universities in Mexico.

The Centro de Investigacion Cientificia y Educacion Superior de Ensenada (CICESE) is the series of buildings on the hill to the left of the highway. This school also has a good reputation in Mexico and the United States.

107 This is the intersection of the alternate route to Ensenada (by Calle 10a) and the coastal route to Ensenada. Geologically the coastal route is more interesting to follow. It provides beautiful views of Punta Banda and Islas de Todos Santos as you drive the last three kilometers along the coast into Ensenada.

109.7 This roadcut exposes a massive andesitic breccia. The metavlocanic rocks in this area are metamorphosed volcanic flows, pyroclastic rocks and sedimentary rocks of volcanic derivation. The most common rock types were basic to intermediate tuffs and breccias, basaltic and andesitic flows, and volcanic graywackes. The age of the rocks in the Ensenada area is still in doubt. Similar rocks range in age from Late Jurassic to the north in San Diego County to Early Cretaceous south of Ensenada. Here they were quarrying the rocks to use to enlarge the harbor. During the quarry operation some of the cliff collapsed onto the highway, so a detour was built around it.

109.8 **ENSENADA** has been a natural port for a long time and blossomed into a major port during the west coast shipping strike in 1974. The strike was a boom to this city and resulted in the enlargement of the harbor. With this enlargement came more business even after the end of the strike. The availability of large blocks of metavolcanic rock in Chapultepec hill has made the task of building the Mulle (pier) and enlarging the port easier and cheaper than importing rock from a more distant quarry. Recent high tides and large waves have demonstrated that the rocks may not be large enough to hold after occasional violent storms which lash the area.

110 This is the intersection of Ave. Gastelum and Calle Primera in Central Ensenada. Follow Mexico Highway 1D (Blvd. Grae L Cardenas) southward along the Malecon on the right. The main tourist shopping district, the destination of many American tourists, is one block to the east. There are a number of good restaurants and hotels in Ensenada. My current favorite is Nico Saad's San Nicolas Inn with its olympic pool. The rooms on the courtyard are the quietest in town.

LAS TRES CABEZAS: The monument on the Malecon commemorates three famous figures of Mexican history: **Benito Juarez**, a great president of Mexico and a full-blooded Indian; **Miguel Hidalgo**, the priest who nailed El Grito to the church door which declared Mexican independence from Spain; and **Venustiano Carranza**, the writer of the Mexican Constitution.

At the south end of the Malecon (Boulevard General Lazaro Cardenas) the highway jogs several blocks to the left on Ave. General Agustin Sangines, passing a large hospital and social services center. Then it turns right on Avenida Reforma (Highway 1) which is the main highway to the south. In 1997 Lazaro Cardenas had been extended one block to bypass the busy Agustin Sangines.

Log 2 - Ensenada to San Quintin [196 kms = 122 miles]

The Highway travels south on the Pleistocene terrace bordered by metavolcanic hills to the left and views ahead of the metavolcanic spine of Punta Banda and the Agua Blanca Fault Zone. It then drops into the fertile alluviated Maneadero Valley and climbs back onto the terrace before turning inland up a valley through the metavolcanics to generally follow along the trace of the Agua Blanca Fault Zone. The highway crests a small pass where it crosses the Agua Blanca Fault Zone, and drops into and follows Santo Tomas Valley which is developed in the metavolcanics along the Santo Tomas and Agua Blanca fault zones. Santo Tomas and the grade south of town are in the Santo Tomas Fault Zone.

The Highway follows a series of oak woodland valleys through more metavolcanic ridges with alluvial valleys and tonalite outcrops to enter the rolling mixed granitic hills and gentle alluviated valley areas of the San Vicente Plain. Beyond the San Vicente Plain the Highway descends a grade and turns seaward through metavolcanic hills to Colonet on the edge of the San Quintin Plain.

At Colonet the Highway turns south to parallel the ocean and follows the Pliocene-Pleistocene marine terrace developed on marine sedimentary rocks of the Rosario Formation and the Pliocene Cantil Costero Formation. The high hills to the east are part of the metavolcanic Alisitos Formation. They formed the shoreline during the development of the Pliocene-Pleistocene terrace. The Highway will alternately climb onto the terrace and drop into alluviated valleys. South of Vicente Guerrero the Baja Highway drops onto the lower late Pleistocene terrace with views of Laguna Figueroa, San Quintin Bay and the basalt cones of the San Quintin Volcanic Field. Near San Quintin the higher marine sedimentary terrace forms a series of mesas backed by the high metavolcanic hills.

0 This log and the kilometer measurements start at the shopping center at the intersection of Calle Gral. (General) Agustin Sangenes and Ave Reforma - Highway 1.

6 The highway heads south along Ave Reforma, parallel to the coast, on the wide and well developed 10 m. Late Pleistocene terrace. In front and to the right are views of the spine of Punta Banda and of the active coastal dune front which is partially stabilized by vegetation of the Chaparral region of the Californian Phytogeographic area.

The **metavolcanic hills** to the left consist of a well preserved section of pyroclastic rhyolite and dacite in a large syncline which plunges to the west. The stratigraphic section exposed here consists of 2,400 meters of lithic and crystal dacite tuffs, breccias, welded tuffs; welded lithic tuff, and andesite; tuffaceous andesite breccia and sandstone, and tuffaceous andesite breccia. A quartz diorite pluton limits the structure and stratigraphy on the north near the Cementos California plant, while a granodiorite pluton terminates the section on the south near the small village of Maneadero. No fossils have been found in this section, but its lithologic characteristics indicate that they are representatives of the early Cretaceous volcanic episodes which have been dated to the south of the Agua Blanca Fault.

12 Ensenada's military airport is for both military and commercial flights. Commercial flights can also be obtained here on cargo planes which service canneries in parts of the Vizcaino Peninsula.

13 The side road to the right leads to Estero Beach Resort on Playa Esteros.

16 The highway leaves the Terrace and descends through Pleistocene dune sands into the fertile Valle de Maneadero. As the highway crosses the Valle de Maneadero there is a view of the hills of Punta Banda. The Agua Blanca fault line, the terraces, the offset streams and the vegetation lineaments show where ground waters have risen to the surface along the faults of this area. The highway approaches the Agua Blanca fault line and will follow it up the valley to the left.

18.5 The side road to the left leads to San Carlos Hot Springs and resort (20 Km.) situated on a fault zone in the metavolcanic rocks.

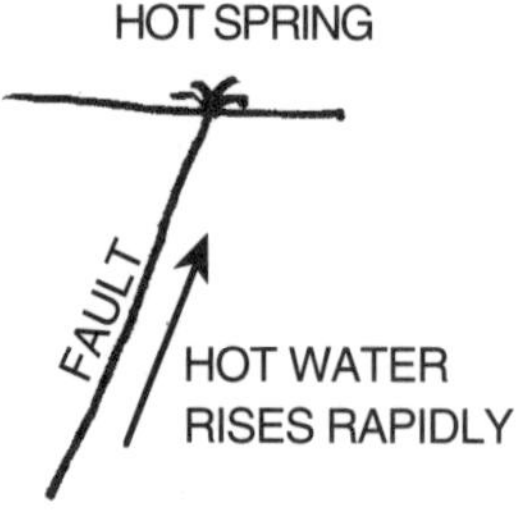

Hot Springs: A hot springs is at least 10° C warmer than the mean air temperature. Hot springs do not normally require abnormal heat sources. The earth's temperature increases about 1.8° Centigrade per 100 meters of depth. Subsurface ground waters are heated to the temperature of the surrounding rock so the 10° warmer water need only come from about 550 meters below the surface. Faults provide an avenue for the rapid rise of hot ground water to the surface which results in hot springs. Most of the hot springs in the western United States are along faults. Some such as Yellowstone and Lassen are near volcanic centers and are due to magma near the surface.

22 This is the turnoff to Punta Banda and La Bufadora (21 kilometers). **The main highway veers left to leave the valley and climbs onto the coastal terrace. To continue south *See* Km. 23 below.**

SIDE TRIP TO LA BUFADORA AND PUNTA BANDA:

The side road turns right towards the coast and makes a number of right angle bends to avoid crossing cultivated fields. As the road approaches the hills, the trace of the Agua Blanca Fault Zone is obvious at the break in slope between the steep hills and the gentler fans.

7 The buildings and trees on the right side of the road are part of the La Grulla Gun Club. The north edge of the slough beyond the club is a scarp and the slough is a sag pond on the Agua Blanca Fault. The road crosses this scarp about 1/2 km to the southeast. Many more subdued fault features can be seen along the base of Punta Banda Ridge ahead (*See* fault figure 2:24)

7.5 At the highway curve, rushes (*Junco sp.*) are growing in on the left side of the road. Numerous resident Red Winged Blackbirds are seen building nests here in the spring (*See* 1:69.2).

8 Castor Bean, Junco, Typha, composites, and Giant Reed are found in roadside ditches. Tamarix and olive trees form wind breaks. The area is an "estero" or salt marsh. Other distinctive plants in the area include Giant Reed (*Arundo donax* of the Toyon grass family), California Pepper Tree, Indian Tree Tobacco, Laurel Sumac, Rabbit Brush, and Datura.

9 The sand spit separating the estero from Bahia Ensenada can be seen ahead to the right. Numerous hot springs along the recent break of the Agua Blanca Fault lie at the edge of the estero a few hundred feet to the right of the road. The western most hot spring is on the main beach and faces the ocean near Km 11 almost opposite the northernmost house. It is possible to dig in the sand here at low tide and encounter hot water (*See* 2:18.5).

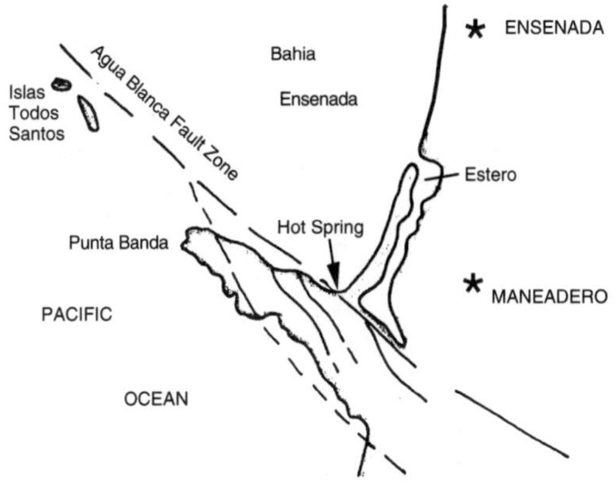

10.6 Turnoff to Punta Estero and the Baja Beach and Tennis Club on the sand spit of Punta Estero. The Pleistocene terrace materials are exposed in the roadcut. Pickleweed, Alanofria, and other halophytes grow in the sands of the estero.

The dunes are stabilized with Wild Buckwheat, Beavertail Cactus, Ice Plant, and annual sunflowers. Only halophytes grow in this region. Datilillo is planted around residences and gate houses which gives more of a tropical flavor to the environment. Annual grasses take advantage of the highway edges.

The avifauna of the mudflat near the mainland in the upper part of the estero consists of Dowagers, Godwit, Cattle Egrets, Western Gulls, California Gulls, and Turkey Vultures. The birds dig for invertebrates, clams, etc. in the mudflats.

11 The vegetation on the hills is typical of the chaparral flora and consists of Cheese Bush, Toyon, Wild Buckwheats, Black Mustard, Locoweed, Tamarisk, and Mission Cactus.

Some of the roadcuts are in faulted and deformed dacite porphyry. This rock is part of a metavolcanic mass that is the backbone of Punta Banda. For the next three kms. the road traverses Pleistocene and upper Cretaceous rocks, faulted against or deposited on the older metavolcanic rocks on the north side of the point.

12.5 La Jolla.

13.2 View to the left rear of the trace of the Agua Blanca Fault. The fault passes through the notch in the skyline and follows the base of the steep slope.

14.2 Cretaceous age rocks were first recognized and described in Baja California at this location. The road into the gully past the houses on the right leads to a small quay on the beach. The Cretaceous rocks are exposed in the sea cliffs along the beach near the quay and in road cuts for the next two Kms.

CORALLIOCHAMA ORCUTTI LOCALITY

Coralliochama orcutti: C.A. White (1885) described *Coralliochama* (aberrant clams) bearing rocks on Bahia Todos Santos. This genus is one of the most characteristic elements in the late Cretaceous near-shore faunas of North-western Baja California. Accumulations of specimens, some with shells attached and partially abraded, can be seen in these sea cliffs along the north side of Punta Banda near the village of El Rincon. The upper Cretaceous beds at this locality are more highly deformed than at any other locality known in this region. This is due to their proximity to the Agua Blanca Fault which lies just offshore.

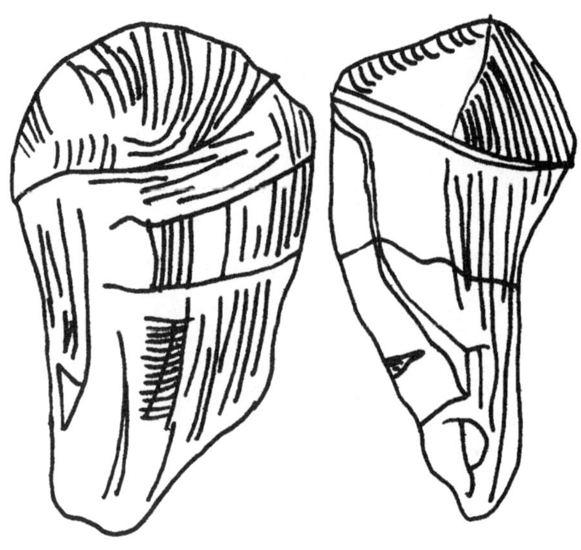

CORALLIOCHAMA ORCUTTI

16.5 Gray-green rocks of the Alisitos Formation are exposed along the road.

The ground at this turn out is covered with Ice Plant (*Mesembryanthemum sp.*) a fleshy leaf succulent naturalized from south Africa. Ice Plant is locally seen on sea cliffs, dunes, beaches, and sandy soils.

17.8 The road crests a grade and briefly passes along the spine of Punta Banda. The Parry Buckeye grows in arroyos and on the hillsides.

18 The village of La Bufadora is below the road to the left. Several Pleistocene marine terraces are obvious along this segment of the point.

The vegetation on the terrace consists of Toyon, Wild Buckwheat, Sumac, Chamise, Agave, Jojoba, Broom Baccharis, and Indian Tree Tobacco.

20 This area of Punta Banda is composed of massive highly jointed metamorphosed andesite porphyry. The blowhole is developed along a large joint in the metavolcanic rocks.

The contact between the andesite and the rounded boulder conglomerates in the Pleistocene terrace is exposed along the road.

21 **La Bufadora:** Trailer parks, campsites, cafes, and numerous small curio shops are located on Punta Banda. These establishments have developed to serve the visitors who come to view the activities of the world's largest blowhole.

LA BUFADORA BLOWHOLE

THE LA BUFADORA BLOWHOLE is reputed to be the largest blowhole in the world. Waves approaching the shore of Bahia Papalote trap air in a rocky cave along a large fracture on the northern shore of the bay and block the cave mouth. As the wave crest moves into the cave, it compresses the cave air until the release of pressure along cracks in the roof of the cave forces a plume of water skyward. The steepness of the wave front and the amount of water blocking the cave entrance determine the strength and height of each plume of water. If there are not many swells, the blowhole may not put on a show.

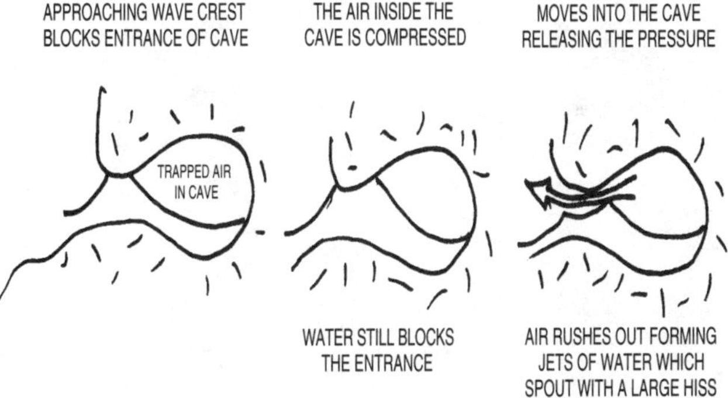

Another attraction can be experienced by divers just offshore in the form of hot springs that have developed along the Agua Blanca Fault. Some divers have hard boiled eggs in the hot waters which rise out of the spring.

From La Bufadora you return to the main highway at Km 22. You may turn right and continue to the south or return to Ensenada.

CONTINUATION OF MAIN LOG

23 An immigration station is located at Maneadero, but it is not always open. In early 1997 this station had not been open for several years. Proof of U.S. citizenship may be required. A certified copy of a birth certificate or a Passport is acceptable, and a single-entry tourist visa must be filled out and stamped by immigration officials on duty. The process is simplified if you have already filled out the slip which may be obtained where you bought insurance.

24 There is a view to the north of the Miocene basalt mesas, Islas de Todos Santos, Bahia de Todos Santos, the point of Punta Banda, and the long spine of hills which comes southeast from Punta Banda along the Agua Blanca Fault Zone. The Diagram (from Borcherdt, 1975) depicts some of the fault features that you are viewing.

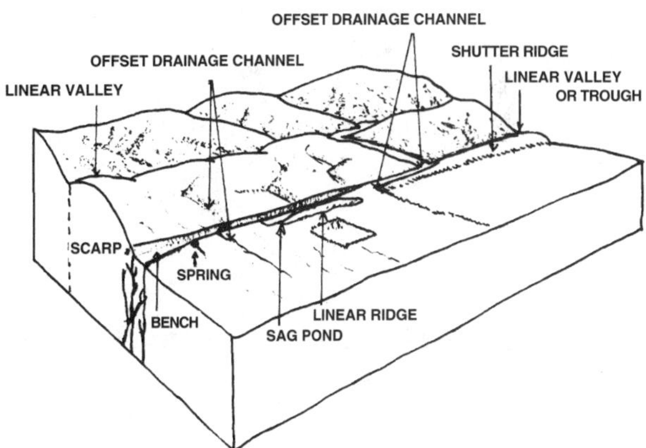

Notice the long straight nature of this spine, the flats along the spine, and the lines of vegetation. Near the middle of the spine and close to the highest point, there are a series of houses on a flat terrace. These houses sit directly on the Agua Blanca Fault Zone. Note the alluvial fans coming down from the fault zone. On either side of the houses there are offset streams which indicate right lateral movement. To the left is a notch in the hills through which the highway will eventually cross into a valley occupied by the Agua Blanca Fault Zone.

33 As the highway follows Canon las Animas, it parallels the Agua Blanca Fault Zone which is located to the right. On the ridge above the highway are examples of shutter ridges, benches, vegetation lineaments, and offset streams which typically characterize a major fault zone.

Age of the metavolcanic strata: There are numerous lower-upper Cretaceous fossil dates in the Alisitos formation along the peninsula south of the Agua Blanca Fault Zone. North of the Agua Blanca Fault Zone there are no lower-upper Cretaceous fossils in the metavolcanic strata. The next fossiliferous rocks are the Santiago Peak Volcanics in the San Diego area; they contain uppermost Jurassic fossils (Fife *et. al.*, 1967). Between the Agua Blanca Fault Zone and San Diego County, there are no fossil or radiometric dates on the metavolcanic strata.

39 Black-shouldered Kites (*Elanus caeruleus*) are commonly seen flying, hovering, and hunting along this stretch of highway.

Black-shouldered Kites or Milano Coliblanco are large falcon-shaped gray birds with black shoulders, a white underside, and a long white tails. They may be seen flying, soaring, or hovering like a kite along the Highways as they hunt small reptiles, rodents, and large insects. Kites are uncommon residents within the coastal plains and valleys from San Quintin north, and occasionally south to Guerrero Negro.

40 The Agua Blanca Fault Zone runs diagonally across the field of view from the right rear to left forward in the middle part of the hills.

45.3 As the highway crests at a small pass, there is a sweeping view of the fault line valley of Valle Santo Tomas.

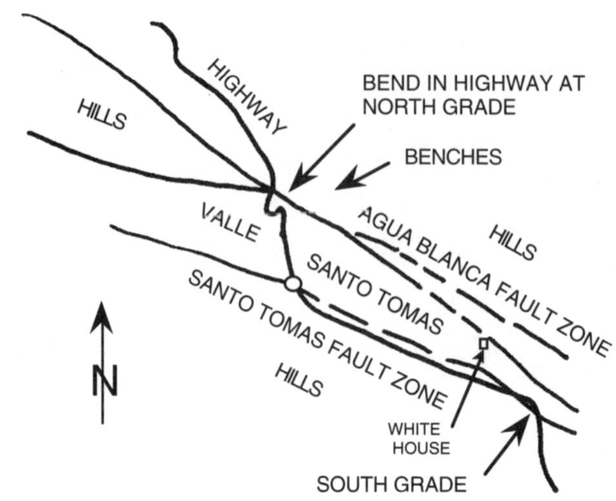

Between the sharp curve to the right at the top of the grade and a hairpin curve to the left (at Km. 46.1), numerous shears and crushed rock provide evidence of the Agua Blanca Fault Zone that crosses the highway near the top of the grade.

47.8 The graded road to the right leads to La Bocana and Punta China (26 kms.) where the upper Cretaceous Alisitos Formation was first defined by Santillan and Barrera, (1930) for the rocks near Rancho Alisitos. A cement quarry on Punta China has been developed in the limestones of the Alisitos Formation

ALISITOS FORMATION AT PUNTA CHINA
EDWIN ALLISON LECTURING

This graded road is the newer higher road that has been cut into the side of the hills to avoid the frequently flooded and washed out older road of the valley bottom. There is a developed camping spot 5 Kms. down this road in the old Oaks and Sycamores at the old Mision Santo Tomas.

50.5 The highway crosses the broad faultline valley of Valle Santo Tomas. The large river bed of the meandering Rio Santo Tomas usually contains only a small perennial stream except during periods of heavy precipitation when it may become a large raging river. Some of the waters of this river probably originate as freshwater springs from subsurface groundwater which rises upward through the crushed rock produced by movement along the Agua Blanca Fault Zone. A Riparian Woodland borders the Rio Santo Tomas.

This water source once supported the small Dominican Mision Santo Tomas and was the reason for the mission's location here. An expedition of Portola and Father Serra named this valley "Canada de San Francisco Solana". In June, 1794, Father Loriente founded a second mission, the Mission of Santo Tomas de Aquino, after abandoning the first mission 5 Kms. further downstream closer to the Pacific Ocean. The ruins of the second mission are behind a house near the highway. Take the new road or the dirt road on the north edge of town to find the first mission. This old road down the valley eventually reaches Punta China and La Bocana over a rougher but more scenic route.

RIPARIAN WOODLAND AT MISION SANTO TOMAS

Riparian Woodlands are vegetational communities which grow along streams and other drainage ways. Since the climatic regime over much of Baja is an arid one, the local occurrence of permanent standing or running water has a dramatic influence on the composition and quantity of vegetation. Localities with permanent standing or running water are generally bordered by large deciduous trees, shrubs, and herbs which only grow on the banks of such water courses. Where river valleys are broad, the Riparian Woodland is correspondingly broad; at higher elevations where the water courses are narrow and the stream banks are steep, Riparian Woodlands may form a very narrow strip which may be only a few meters wide.

COASTAL SAGE SCRUB OF CALIFORNIAN PHYTOGEOGRAPHIC REGION

51 **SANTO TOMAS:** The small village of Santo Tomas is located in the green fertile Valle Santo Tomas. This old agricultural community dates back to mission times. It was the original site for the famous Santo Tomas Winery. However, the winery was eventually moved to Ensenada and today olives are the main crop of Santo Tomas.

The El Palomar restaurant is directly in the Santo Tomas Fault Zone. This fault forms the south side of the valley and the Agua Blanca Fault Zone forms the north. To the north and south of the valley the steep hills are densely covered by Coastal Sage Scrub.

Looking at Baja's faults: Sit on the white rocks at the outside edge of the shaded parking area across from the gas tanks and look at the hills on the north side of the valley. On your left side to the north, the Agua Blanca Fault passes through the same low pass as the highway. As you turn to the right, the trace is marked by a series of benches about 1/3 of the way up the hills from the valley floor. These benches descend to the valley bottom at the major stream arroyo which is almost directly across the valley. There they give way to a valley which parallels the main valley. To the right the fault follows the notch between the two low yellow-brown hills which are parallel to the axis of the valley, and behind the far right ridge.

58 The single-story white house across the valley at the base of the hills on the left marks the site of a spring which is located on the main trace of the Agua Blanca Fault Zone.

57 Numerous small fans which mark a trace of the Santo Tomas Fault have developed at the base of the hills to the right. Due to relatively recent uplift along the faultline, streams are cutting into the small fans.

60.1 The highway climbs the grade where roadcuts exhibit numerous sheer and crushed rock zones typical of the trace of a major fault zone. The turn-off to the left provides a good valley view.

AGUA BLANCA FAULT VALLEY

61 The turnoff near this kilometer marking provides a good view of the Agua Blanca Fault Zone which extends through the valley to the north. There is a line of bushes and very small trees which cut across one of the points just above and to the right of the white ranch house on the opposite side of the valley. This vegetation line is replaced by the low notch to the west where the crushed rock of the fault zone was easily weathered.

The northern end of the Agua Blanca branches near here to form the north and south sides of Punta Banda. Between these two branches a resistant sliver of metavolcanic rocks of the Alisitos Formation forms the spine and point of Punta Banda.

63 At the crest of the highway, there is a well-graded side road to the left which leads to a microwave tower on Cerro el Zacaton. This dirt road crosses the Agua Blanca Fault Zone.

68 The area to the left is Valle San Jacinto. The side road to the right is one of several which lead to Punta Cabras and Punta San Jose (to the northwest) and Erendira (to the southwest) The preferred route is at Km. 78.5. Both of these areas have beautiful sea cliffs and pocket beaches.

SEA CLIFFS AND POCKET BEACHES: There are many beautiful coves and sandy pocket beaches with sea cliffs of Pleistocene terrace material overlying the Cretaceous Rosario Formation. They resemble the beautiful beaches at La Jolla, California. Generally the sea cliffs in this area are formed on the Cretaceous Rosario or Alisitos formations. The pocket beaches form when the sea cliff is capped by softer marine terrace deposits. The fractures in the harder cliffs are enlarged into coves by the waves. This leaves a cove with a pocket beach, which usually consists of coarse sand and gravel.

69 View to the left rear. A series of light-colored exposures run along a bench in the middle of the skyline ridge. These light-colored exposures are the crushed rocks which mark the trace of the Agua Blanca Fault Zone.

73.3 The metamorphosed volcanic and sedimentary rocks of the Alisitos Formation are especially well exposed in the roadcuts along this arroyo. Some of the holes in the rock are the impressions of the shells of gastropods which have been leached from the rock.

The streams and valleys in this area exhibit Riparian Woodland species of Oaks, Willows, and California Sycamores, while the hillsides contain the typical Coastal Sage Scrub vegetation represented by Brittle-Bush, Laurel Sumac, Broom Baccharis, Chamise, Wild Buckwheat, and Dodder or "Witches Hair" growing on plants.

75.5 The resistant ridge, which crosses the highway at this bend near La Angastura, is formed by a series of light-gray limestones within the Alisitos Formation. These limestones are quarried at Punta China by a commercial concer, Cementos California. The limestones, which are low in magnesium and aluminum, outcrop in a belt which extends southeast from Punta China. The large reserves and their proximity to deep-water shipping facilities make it an important source of limestone for cement.

78.5 A paved road marked "Erendira" leads south 6.6 kilometers down Canon Guadalupe and Canon San Isidro to Arroyo San Vicente, and then turns to the west to the small coastal community of Erendira and Bocana Erendira.

80.5 The highway crests over a ridge with a panoramic view of the San Vicente plain. Mision San Vicente Ferrer is located on the right before the bridge. This plain is dry farmed for wheat and barley.

Near San Vicente the metavolcanic rocks of the Alisitos Formation give way to tonalite. Much of the San Vicente plain is underlain by tonalite and a mixture of metamorphic and plutonic rocks.

89 San Vicente originated in 1780 as a Dominican mission site, but it fell into obscurity until the 1940's. Rejuvenation of the community was the direct result of agricultural development in the area.

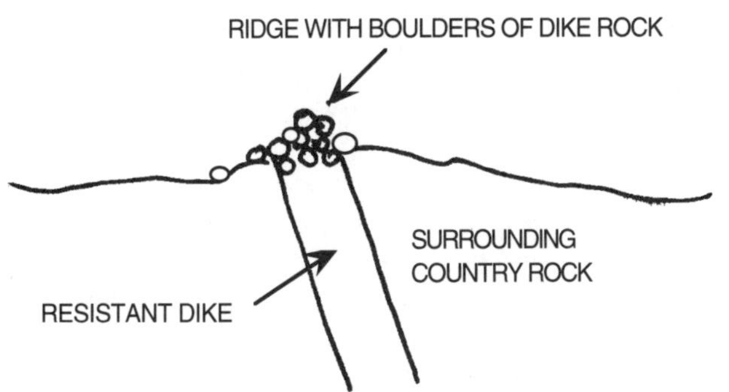

93 For the next kilometer the low linear mounds of light-colored boulders are the weathered remnants of resistant granitic dikes which have filled shrinkage cracks near the top of the batholith. The surrounding country rock (tonalite) is less resistant and more easily weathered and eroded. This leaves the ridges of dike rock.

102 This flat plain is Llano Colorado (red plain). The view to the east is of the metavolcanic foothills.

Valle Llano Colorado supports the most extensive grape vineyards and olive orchards in Baja.

107.5 The highway crosses an arroyo with dense Saltbush plants of the genus *Atriplex*.

Saltbush is a halophyte which has modifications which enable it to grow in very salty dry soils. The seeds of the Saltbush contain chemical growth inhibitors which dissolve when exposed to sufficient amounts of water. These inhibitors ensure that the seeds will only germinate when sufficient quantities of water are available to allow the seedling to complete its life cycle.

Saltbush can grow in soils with a high salt content. The salt is absorbed into the plant through its roots. The leaves can be used as a flavoring for foods, and the parched ground seeds make a tasty coarse meal or fine flour.

109 Granitic dikes are exposed along the grade.

115 The highway descends through granodiorite and gabbro outcrops into Arroyo Seco where the highway turns to follow the arroyo toward the sea. Metavolcanic rocks of the Alisitos Formation are exposed along the upper Arroyo Seco.

This point was the end of the paved highway for many years. Until the completion of Highway 1 the sign read **"AL FIN DEL PAVMIENTO"** (to the end of the pavement).

The roadbed was prepared all the way to San Quintin but was not paved. "World class" potholes were encountered on this graded segment. Most of the traffic made new "roads" alongside the graded segment in an attempt to avoid the potholes and ruts.

123.5 There is a bakery in the brown house on the right. They usually bake for the evening meal, so the selection in the early morning is often limited. The "beehive" oven (horno) is visible from the back of the building.

126 This Gas Station and restaurant was once run by Wilmont Bradley, a Baja pioneer who died of cancer. He built the medical clinic on the hill. For many years this clinic was the local center for the Flying Samaritans. A group of doctors still fly into Mexico to set up clinics which provide medical care that is often otherwise unavailable

OLD ROAD AND ITS CHOICES

126.4 The highway crosses Arroyo Seco. The side road to the right just before the bend in the highway leads to San Antonio del Mar which is located on a beautiful sandy beach.

126.6 The highway passes through Punta Colonet.

128 Look to the right rear for a view of Punta Colonet. This point was named after an English sea captain who for some unknown reason ran his ship aground on this treacherous coast. The survivors walked to Ensenada.

130 The highway dips into a small steep gorge whose rocky cliffs are the metavolcanic rocks of the Alisitos Formation.

131.5 The low range of hills, west of the highway, are stabilized coastal sand dunes.

132 The red clay soils in this area are derived from the iron-rich Alisitos Formation. The Llanos de San Quintin is drip irrigated and dry farmed.

133 The highway begins to follow a flat 30 meter Pleistocene terrace which is often called the Llanos de San Quintin.

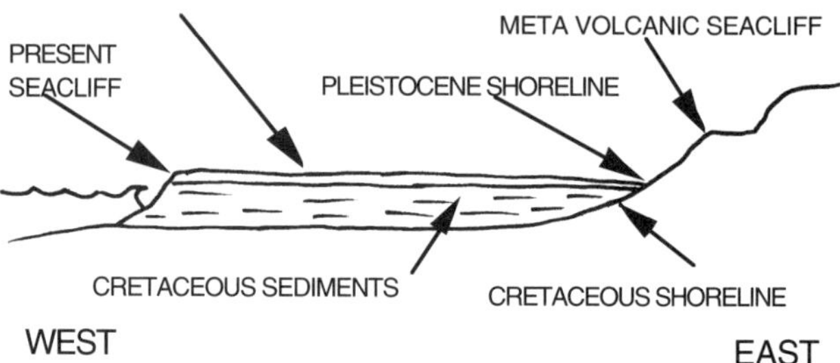

To the left are the foothills of the Alisitos Formation. The shoreline of the ocean during the Cretaceous, when the terrace rocks were deposited, and during the late Pleistocene, when this terrace was cut, is roughly equivalent to the edge of the terrace and the base of the low foothills.

140.9 The road to the left leads to the small village of San Telmo (10 Km), the Meling Ranch (50 Km), Mike's Sky Rancho, the pine groves of San Pedro Martir National Park, and the new National Observatory (87 Km).

Plutonic rocks of the Peninsular Ranges: Reconnaissance mapping in the Peninsular Ranges has resulted in the identification of 387 plutons over 1 km in diameter which cover 28,000 square miles. These plutons are smaller on the western edge of the range; the axial portions of the range are occupied by a relatively small number of much larger bodies. Most of the plutons are circular in outline and many show concentric structures. Granite and gabbro form the smallest plutons, while most of the largest plutons are composed of tonalite. These plutons yield isotopic Uranium age dates of 95 to 119 million years before present (MYBP). K/Ar dates of 75 to 85 MYBP in the western edge and in the

60's on the eastern edge yield cooling dates or ages of metamorphism (Gastil *et.al.*, 1975).

The rocks on the road to the Meling Ranch and the high San Pedro Martir exhibit a cross section of the pre-batholithic and batholithic rocks of the Peninsular Ranges. The road begins in the metavolcanic and metasedimentary rocks of the Alisitos formation. This area is stratigraphically the lowest strata in the type area of the Alisitos Formation. A few kilometers up the road there are the strata which contain Albian fossils (Allison, 1974). Nearer the ranch these rocks are progressively more metamorphosed until they become schists and gneisses. At the ranch a pluton of granitic rocks is exposed. As the road climbs to the observatory, it enters an extensive area of granitic rocks.

Flora of the Sierra San Pedro Martir. The Sierra San Pedro Martir averages 1,500-2,800 meters or more in elevation. Its western slopes are covered with coniferous forests which receive 50 centimeters of precipitation per year on the average. The dominant species of the coniferous canopy are Jeffrey Pine, Incense Cedar, Sugar Pine, Lodgepole Pine, and Colter Pine. At lower elevations Quaking Aspen and Canyon Oak frequently occur. The understory of the forest consists of shade-tolerant annuals, grasses and widely spaced shrubs.

At elevations below 1,000 meters the western slopes of this range is clothed by true Chaparral and Coastal Sage Scrub. The drier eastern slopes are sparsely covered with vegetation which is characteristic of the San Felipe Desert, a southern extension of the Colorado Desert of the United States. (The vegetation of the San Felipe Desert will be discussed at 14:108).

At the southern end of the Sierra San Pedro Martir the flora is in transition between the Chaparral vegetational area of the Californian Phytogeographic Region and the desert flora of the Desert Phytogeographic Region. As a result, the southern slopes are covered with a mixture of northern chaparral species and the desert vegetation of the Vizcaino Desert. This region of overlapping floras is called an ***ecotone*** and extends approximately 160 kms. from Valle Santo Tomas, to as far south as El Rosario. This zone contains mixed Chaparral and desert flora which grade into each other at the respective ends.

141.3 A road metal quarry has been developed in the Alisitos Formations on the left.

The ecotone vegetation of this region consists of a mixture of the following Chaparral, Coastal Sage Scrub and desert plant species: Candelabra Cactus, Wild Buckwheat, Burrobrush, Burr Sage, Agave, Jumping Cholla, Brittle-Bush, Beavertail Cactus, and annual composites.

151.5 The highway climbs through more upper Cretaceous Rosario Formation outcrops. Pliocene fossils have been collected from the Pliocene Cantil Costero Formation in the second and third quarry at Km 155.5.

This area is an uplifted fault block called a "horst". Erosion of the uplifted area causes the exposure of the sedimentary rocks.

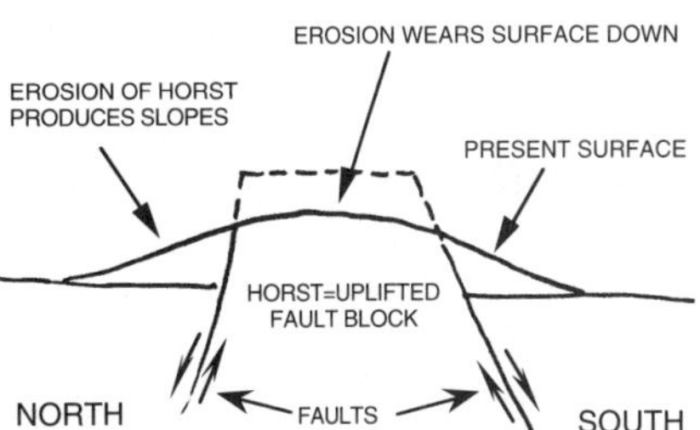

153 Some of the hummocky topography on the horst is probably due to stabilized coastal dunes. These dunes are much more obvious on the next rise to the south.

154.3 Pliocene fossil localities are located in the road metal quarries.

156 The highway begins a descent and passes through metavolcanic outcrops and then through Pliocene outcrops on the south side of the horst.

157 Camalu

165 Some of the roadcuts in this area are composed of dune sands. The rolling hills in this area are a stabilized coastal dune field that is extensively plowed and planted with various agricultural crops. This could lead to the destruction of the fragile dune environment and to the development of "dust bowl" conditions if the plowing and planting is not done properly and at the correct time.

DUNE STABILIZATION: The stabilized dune systems of Baja provide a delicate environment that is home to numerous plants and animals. Farmers take advantage of the retained moisture to plow the dunes and grow crops. However, plowing eliminates the vegetation that stabilized the sand, and winds may blow away the dune. Dry farming and the environment provided by stabilized dunes for the support of plants and animals may be eliminated. For these reasons care should be taken in developing stabilized dunes for agricultural purposes.

It has been noted that there is an increased amount of sand which blows across the road. This is due to the disruption of dune stabilization caused by farming practices. Remember the U.S. Dust Bowl of the 30's.

169 This turnoff leads east 4 kilometers to the ruins of Mision Santo Domingo. This mission has been abandoned since 1885.

169.8 The highway crosses Arroyo Santa Domingo. This bridge was washed out in early 1978. It was rebuilt and a second one was added to accommodate future flood waters.

The hills to the left of the highway are composed of metavolcanic and metasedimentary strata of the Alisitos Formation.

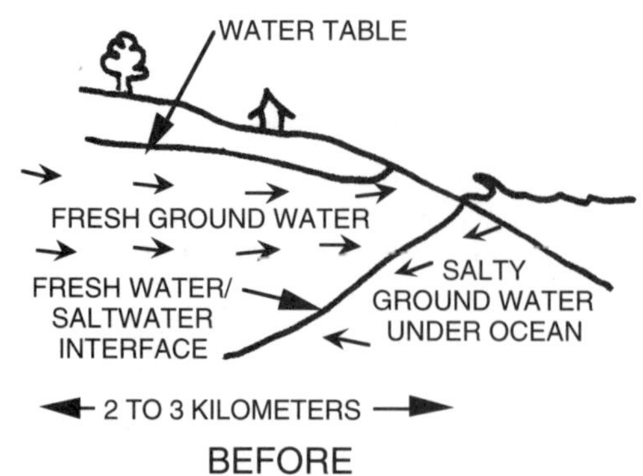

BEFORE

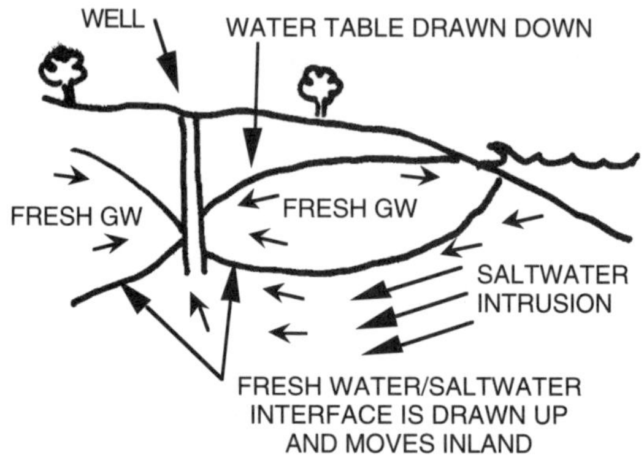

AFTER

1978 ARROYO SANTO DOMINGO WASHOUT

170.2 Vicente Guerrero.

The San Quintin plain is extensively dry farmed and irrigated from groundwater wells. The over usage of the groundwater near the bay has resulted in salt water intrusion; the salty groundwater under the bay pushes inland (intrudes) and displaces the fresh water which is withdrawn.

176 From the crest of this hill there is an excellent view of the San Quintin volcanic field.

SAN QUINTIN VOLCANIC FIELD

Approximately ten small cones of the **San Quintin volcanic field** are visible from the highway. To the right of the main group of cones is the silhouette of Isla San Martin, another volcanic cone 4 kilometers offshore. Woodford (1928) was the first to describe the San Quintin volcanic field in great detail. This field is a series of olivine basalt cinder cones and flows of Pleistocene to Recent age. Locally, marine terraces are developed on the lava flows. This indicates that the eruptions predate the falling sea level and the 125,000 year old Sagamonian Terrace. As the highway descends to traverse Llanos de San Quintin, the shallow Laguna Figueroa is visible behind the dunes along the coast of Bahia San Ramon. This area has been exploited for salt since mission times.

MEDIA LUNA CINDER CONE

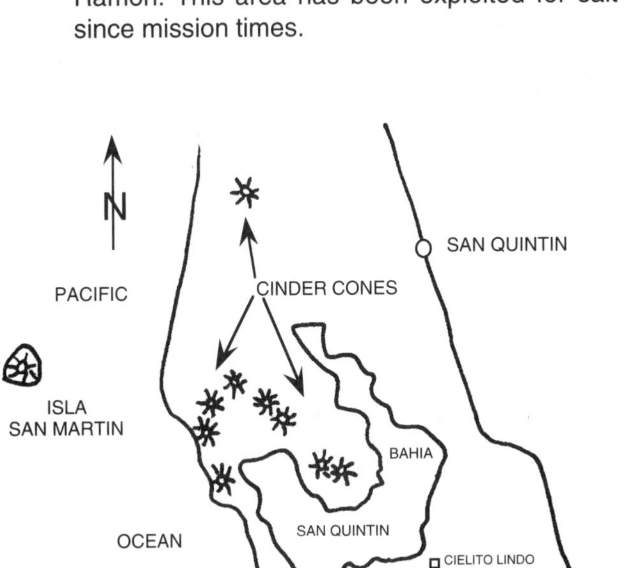

183.1 I was involved in an automobile accident near the store. All premiums on Mexican auto insurance were returned when we slid into a truck which made an illegal left turn. Be sure to have Mexican Insurance!

187 The hills to the left consist of the Cretaceous Rosario Formation.

The hill to the right is Volcan Media Luna (half-moon), one of the basaltic cinder cones of the San Quintin Volcanic Field.

189.5 At **San Quintin** a Pemex station (Magna Sin) and stores with adequate supplies, a good auto parts store, mechanics, hotels and banks, and an ice house are available.

191 The bridge at this kilometer marks the south end of San Quintin.

193 Lazaro Cardenas is the site of a military base.

196 The kilometer markers along the side of the road at Km. 196 change to 0.

Log 3 - San Quintin to Bahia de Los Angeles Turnoff [285 kms = 177 miles]

***San Quintin to El Rosario** - South of San Quintin the Highway skirts a dune area and follows the low terrace along the base of the high mesa of Cretaceous marine sedimentary rocks overlain by Pliocene marine sedimentary rocks. The high mesas become much wider in this area following a relatively straight Cretaceous shoreline. The steep metavolcanic hills are out of view to the east.*

At Consuelo the Highway leaves the coastal terrace and climbs up a valley through Cretaceous hills to the high mesa surface with the metavolcanic hills in the far distance to the east. The highway briefly crosses the mesa before descending a steep canyon through the Rosario Formation to El Rosario.

***El Rosario to Catavina** - This stretch of the Highway travels almost directly east. From El Rosario the highway turns inland up the Rio Rosario on a river terrace, crosses the river, and passes through rolling hills of Cretaceous marine sedimentary rocks. The highway climbs up on the rolling mesas with distant views of the resistant Paleocene mesas to the south and the high granitic San Pedro Martir to the north. The highway crosses the Cretaceous shoreline into the rolling metavolcanic foothills and descends a grade into a series of gentle valleys developed in the metavolcanic and granitic rocks. The highway then follows a broad valley with steep ridges of metavolcanic rocks on both sides of the valley.*

After El Progresso and a low pass in the metavolcanic rocks the Highway begins to traverse the San Agustin Plain on lakebed sediments. The steep metavolcanic hills to the left formed the shoreline of the lake. The highway follows a flat to rolling surface developed on fluvial conglomerates with higher hills of granitic and metavolcanic rocks to the south. To the north is the extensive flat area of the San Agustin Plain which is broken by isolated low metavolcanic hills in the western part and volcanic mesas in the eastern part. Passing between several volcanic mesas the highway enters the rolling hills of a picturesque bouldery tonalite area, finally descending through Arroyo Catavina to Catavina.

***Catavina to L. A. Bay Turnoff** - The Highway crosses a palm studded arroyo to Rancho Santa Ynez, then turns south through a rolling area of bouldery tonalite. Along the route the volcanic mesas become more numerous until, at Jaraguay, the highway climbs to cross an extensive undulating basalt plateau dotted with cinder cones. The bouldery hill of Pedregoso rises through this plateau. The highway then skirts the edge of the plateau in mixed granitic and metamorphic rocks and drops into the flat alluvial valley and lakebed of Laguna Chapala with dunes at the south edge of the lakebed.*

After a brief climb into a low pass in the steep tonalite hills the Highway crosses the Peninsular Divide with views of a wide valley and of the rugged Sierra de la Asamblea. The highway drops into the alluviated valley through tonalite and then Miocene fluvial sedimentary rocks to the granodiorlte hill of Cerrito Blanco. After again passing over a very gentle Penninsular Divide, in alluvium, the highway begins to descend the Arroyo Leon drainage through the Miocene fluvial sedimentary rocks to the junction with the highway to Los Angeles Bay.

0 The kilometer markers along the side of the road at Km. 196 change to 0.

1 This well marked turnoff is the best road to take to the Old Mill (Molino Viejo) and Ernesto's Motel on Bahia San Quintin.

Old Mill (Molino Viejo) About 1885 an American company tried to settle the area. They built a flour mill and pier, and tried dry farming the area for wheat. Today modern drip methods are providing for a thriving agricultural industry.

5 The highway takes a wide swing to the left to cross the arroyo over a high bridge near the hills. The hills to the left, which form low bluffs, are part of the Rosario Formation and are overlain locally by the Pliocene Cantil Costero Formation.

Look and listen closely for Western Meadowlarks (*Sturnella neglecta*) along the highway.

The **Western Meadowlark or Triguera de Occidente** "Sabanero" belongs to the same family as blackbirds and orioles which breed in summer in the northeastern part of Baja and winter throughout the peninsula where they are commonly seen and heard in fields and on fences. Physically, the Western Meadowlark is a heavy-bodied medium-sized bird. The most distinctive identifying field characteristic is a black "V" on its bright yellow breast. The voice of the Western Meadowlark is a rich loud and bubbling whistled song that suddenly interrupts the silence of Baja's desert landscape.

WASHOUTS: In the past the main road went straight ahead through the wash. After repeated washouts most of the traffic followed an easier route which crossed the wash at the base of the hills near Consuelo. When the highway was finally paved in the early 1970's, it once again went directly across the wash where it was again repeatedly washed out and finally permanently located nearer the hills.

8 The highway approaches the Rosario sea cliffs where it roughly parallels an ancient shoreline. The highway is built just above what would have been the old shoreline when the Llanos de San Quintin was cut as a marine terrace.

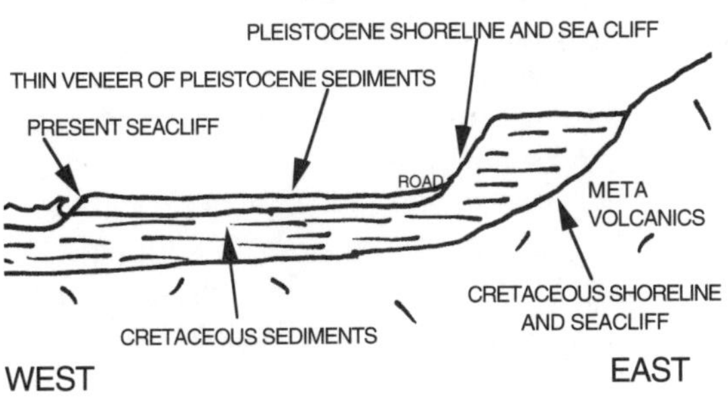

CHANGING SHORELINES: South of Colonet the Cretaceous and Pleistocene shorelines correspond. The metavolcanic hills at the edge of the terrace formed both the Cretaceous and Pleistocene shorelines. South of San Quintin the edge of the low terrace represents only the Pleistocene shoreline. The Cretaceous shoreline has undergone a greater amount of uplift and is on the level of the upper terrace. A climb onto this terrace will yield a view of the Cretaceous shoreline which is fringed by the same meta-volcanic hills that are seen to the north.

8.5 The highway crosses a large arroyo by a high well-built bridge. This used to be the small town of Consuelo. There was a restaurant here, and it was the end of the bus line before the transpeninsular highway was paved.

If you want to see the red bed and mudstone outcrops of the Rosario Formation, park on the left just before the bridge and walk into the road metal quarry about 100 m. from the highway.

10 The Tamarisk-lined side road to the right leads to Hotel La Pinta and Cielito Lindo Motel and RV Park. Juanita, who runs the Cielito Lindo Motel, is one of the best hostesses that the authors have found in Baja. Some say that they have never had a better meal than her rock crab claw plate.

For many years this hotel was the site of an American movie colony retreat which was owned by Mark Armistead, a cinematographer and inventor of the instant replay. The more notable frequent guests were Jimmy Stewart, Henry Fonda, and John Wayne who actually worked on one of the rooms.

TAMARISK TREES (SALT CEDAR): The side road which leads to Cielito Lindo is a very picturesque lane which is arched by two parallel wind rows of old Tamarisk trees which were planted to form fast-growing wind breaks for the agricultural areas of this region. A ride down this road at night is a little eerie.

Tamarisk Trees (*Tamarix pentendra*) were introduced into North America in the early 1800's for use as decorative plants and the formation of fast-growing windbreaks. A single plant may produce 600,000 seeds per year. Its leaves secrete salt, an adaptation to reduce the tree's salt content, which allows it to live in saline soils. Surface soils under the tree accumulate salt which inhibits the germination and establishment of non-salt-tolerant natives. The large quantities of duff (litter) produced by the Tamarisk encourages fire. They can resprout from their own roots following a fire, while many native species cannot. Soils inhabited by Tamarisk soon become arid, since each tree transpires hundreds of gallons of water daily from the soil. These factors aid in the spread of the Tamarisk and reduce the numbers and influence of native species.

BEACH DUNES: At Cielito Lindo the beach dunes are migrating to the south along the surface of the back beach. They are approximately

4-6 feet high and continually migrate to expose material that was covered earlier. During high tides the dunes are often wave-leveled forming a flat surface. As the beach dries, the sand again forms migrating dunes.

WARM SANDY BEACHES AND HIGH TIDES: The warm sandy beaches along Bahia Santa Maria are some of the finest in Baja. They are wide and flat and are covered with sand dollars and Pismo clam shells. At low tide the live sand dollars form interesting and unusual patterns as they move their spines to burrow into the sand. If you are careful, you can cross the dunes and drive on the beach for several kms.

SAN QUINTIN BEACH

SAND DOLLAR TRAIL

Hotel La Pinta used to have a swimming pool and a sea wall with a downstairs bar and restaurant. High tides and tropical storms in 1976 put three feet of water into the lower story of the hotel and filled the pool. The sea wall is gone and the dunes are encroaching upon the area.

HOTEL LA PINTA AFTER STORM TIDE

The brackish water marsh which surrounds the Hotel La Pinta has expanded greatly in the last 10 years. Coots, Cormorants, California Gulls, Western Gulls, ducks, and several species of Grebes are commonly seen in or near the marsh.

The marsh vegetation consists of salt-tolerant halophytic plants such as Sand Verbena and the many branched, jointed stemmed, succulent Pickleweed (*Salicornia pacifica*). This salty plant is edible in salads and can be boiled and used as a pickling mix.

17 The hills of the Pleistocene shoreline converge with the road as the low terrace narrows and disappears under a dune field.

18.5 The highway passes Rancho Las Parritas and climbs onto the Socorro Sand Dunes. The source of the sand is the beaches near Cielito Lindo and Bahia San Quintin (see 3:10). When the migrating dunes reach the curve in the coastline to the southeast, they are blown inland.

20.5 At the crest of the first hill there is a sweeping view to the rear of the San Quintin volcanic field and the flat Llanos de San Quintin.

THE HIGHWAY EDGE EFFECT: In the disturbed soils along the roadside, there are several species of mesic opportunistic plant species. The most obvious are yellow sunflowers, gray Atriplex, Burro Weed and several species of small herbaceous annuals. These opportunists are not seen growing far from the highway.

In the desert regions of Baja the dominant vegetation consists of plant species adapted to long hot dry summers. Dry adapted desert species which require little water are known as xeric plants. (The word xerox means dry and has been used commercially to refer to dry copies produced by Xerox machines as opposed to the wet copies produced by older mimeograph equipment). In a region as rigorous as the Californian and Desert Regions of Baja, micro-climates become particularly important. Any fortuitous or fortunate circumstance which provide habitats a little less xeric, where moisture remains somewhat longer than in more open environments, favors the growth of opportunistic vegetation. Highway pavement edges are classic examples. Concentrations of taller greener and more mesic plants flourish along the pavement edges because additional moisture needed to support more mesic species which are not adapted to the dry desert, is available along the edge of the highway.

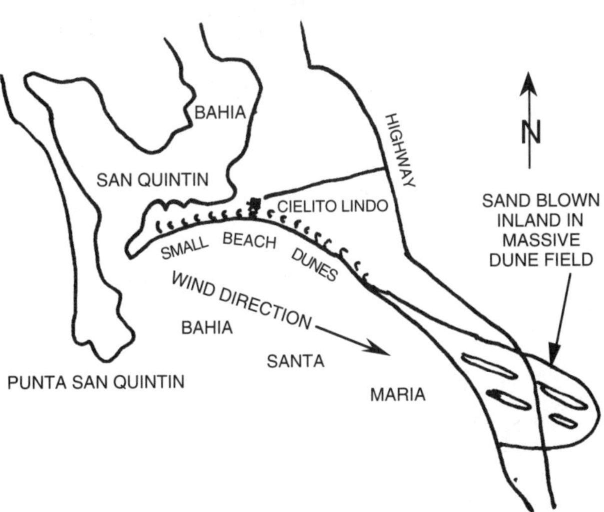

The plants commonly seen on the stabilized sand dunes are *Ephedra*, annual grasses, composites, Bursage, Burro Weed, Jojoba, Ice Plant, and several species of cacti. The dune vegetation visible here is usually less than one meter tall and is fairly dense.

21 In the old days crossing the dunes was very torturous and treacherous. Shifting sands across the road made it very difficult for a 2-wheel drive vehicle. Loads of cobblestones were brought into the area in an attempt to improve these conditions. However, this resulted in a jolting and jarring drive over the cobblestone road. Accidentally driving off the cobbles onto the smooth sand often meant getting stuck.

The view across the dune field shows rows of stabilized longitudinal dunes with unstabilized dunes closer to the beach and the San Quintin volcanic field to the rear in the distance.

SAND DUNES IN BAJA: In the deserts of Baja wind often heaps sand particles into mounds and ridges called sand dunes. They move slowly in the direction of the wind. In Baja dunes are found wherever there is abundant sand and where the wind and the wind direction is fairly constant. All dunes result from local interruptions in the general wind-flow patterns. When the prevailing wind is obstructed by rocks and plants, it is forced to drop its load of sand, building a dune or series of dunes. Once they are formed dunes migrate with the prevailing wind. Sand is transported from windward to leeward where it gathers at the dune foot and constantly encroaches on new territory. Not all dunes travel. Where winds seasonally reverse or are multidirectional, they tend to remain stationary. Vegetation often helps to stabilize smaller dunes, particularly those that have developed around an embedded plant.

Although a sand dune looks like an inhospitable dry environment, it supports a great many trees such as Mesquite, Tamarisk, and Catclaw Acacia; bushes such as Four-winged Saltbush, Creosote Bush, Desert Buckwheat, Golden Bush, and Mormon Tea; and grasses such as Ricegrass and Big Galleta. A number of annuals carpet the dunes during rainy parts of the year. Some of the most commonly seen annuals on dunes in Baja are the Sand Verbena, Evening Primrose, and many species of composites. These trees, bushes, grasses and seasonal annuals enable the dunes to remain where they are by stabilizing the sand with their branches and roots. Although dunes may look dry, they are not. All precipitation that falls on the dunes soaks in and very little runs off. Though the surface dries quickly which gives the impression that the interior is moistureless, water remains in the dunes long after the regions surrounding the dunes are dry. This provides moisture for plant growth.

The dune vegetation provides shelter, food, and protection for numerous animals which include insects, centipedes, scorpions, birds, lizards, rodents, rabbits, gophers, and ground squirrels. The stabilized dune systems of Baja provide a delicate environment that is home to numerous plants and animals.

23.5 Road to old Rancho El Socorro

24 The highway continues to the south over a Pleistocene marine terrace cut into the Cretaceous Rosario Formation. The Cretaceous shoreline is now several kilometers inland. It will be crossed again near Km 80 south of El Rosario. The hills to the left consist of Rosario Formation mudstones in the lower part overlain by Paleocene Sepultura Formation.

However, there are quite a few different kinds of plants growing here. Purple Bush is the dominant plant cover of the flats. There are a few interspersed specimens of Desert Thorn, Velvet Cactus, Agave, Cholla, and Bursage. Tamarisk is the dark-green plant of the wash bottoms, and patches of red-colored Ice plant and young Brittle-Bush grow in the disturbed soils at the pavement's edge.

29 The small hill to the right on the sea cliff is an old stabilized dune which was reactivated as the vegetation is being removed by the erosive action of the sea. Other dunes such as this can be seen at kilometers 36.5 and 41. The sea is cutting into the base of the sand dune which demonstrates that there has been regression of the sea cliff since the formation of the dune.

35 The highway crosses Arroyo Hondo with exposures of the Pleistocene Terrace material.

40 The Cantil Costero Formation was named for the exposures of marine Pliocene rocks which cap the mesas to the left of the highway. Distinct wave-cut terraces seen in this area were cut by the sea over a period of time. The highest terrace is developed on the Cantil Costero Formation (Late Pliocene). The Pliocene shoreline occupied nearly the same position as the present shoreline but has been uplifted approximately 100 meters since it was formed. The cutting of the marine terrace into the Cantil Costero Formation during the Pleistocene has exposed the old shoreline. The lower terrace, where the highway is constructed, corresponds with the Sagamonian high sea level stage 125,000 years ago.

The large gray blocks on the slope to the left are fossiliferous conglomeratic limestone of the Cantil Costero Formation. In many places slightly above the limestone are fossils which include specimens of the large barnacle *Balinas tintinabulum*, which may be 15 centimeters tall. A commercial operation has removed most of these barnacles to be sold in shell shops. Along the seashore a pair of stabilized dunes have been cut by the sea. The southern dune is still partially stabilized.

CANTIL COSTERO FORMATION IN CLIFFS

41.5 The highway turns inland at El Consuelo, and climbs through the subdued hills of the Rosario Formation to the top of the mesa.

This is the last view of the Pacific Ocean for several hundred kilometers.

The vegetation becomes more dense. The plant species commonly seen along the highway are Burrobrush, Pitaya Agria, Agave, Garambullo, Beavertail Cactus, Brittle Bush Nipple cactus, *Ephedra*, Velvet Cactus, Hedgehog Cactus, and occasionally Candelabra, Ice Plant, Maguay, Cliff Spurge, Lichen, and Milkweed. Locoweed (*Astragalus sp.*) is the plant which grows abundantly along the pavement's edge.

LOCOWEED: Many species of Locoweed grow in the Californian Phytogeographic Region of Baja. They are distinguished by the length of floral parts and the size of the wooly seed pods. *Astragalus*, members of the pea family (F. Leguminosae), exhibit a variety of colored flower spikes (purple, pale yellow, and white) and occur in many different types of plant communities. Their leaves are alternately arranged along the narrow stems and are unequally pinnately compound (the leaf veins give the leaf a feather-like appearance).

Astragalus is commonly known as locoweed because it contains an accumulative toxin, Selenium, that is poisonous to practically all live-stock. They repeatedly eat the toxin and go "loco" prior to dying from selenium poisoning. Although it is poisonous to cattle, early Baja Indians chewed the shoots to cure soar throats, used locoweed poultices to reduce swelling, and boiled roots which produced a decoction which was used to wash granulated eyelids and ease the pain of toothaches. "*Astragalus*" comes from Greek and means anklebone, an early name for leguminous plants. Thus, the plants name may have referred to their low prostrate growth habit which placed them at ankle height.

49 The highway climbs a grade toward the top of a mesa through mudstone exposures of the Rosario Formation.

50 At the top of Mesa las Cuevas the vista to the northeast consists of mesas composed of Paleocene rocks overlying the Rosario Fm.

52 The highway descends a grade into Canada el Rosario passing through the sandstones, mudstones, and conglomerates of the Rosario Group.

Rosario Group: Kilmer (1963, 1965) mapped and named four formations of Late Cretaceous age: the Rosario Formation for the upper marine part, the El Gallo Formation for the underlying nonmarine unit, the Punta Baja Formation for a lower marine unit and the La Bocana Roja Formation for a basal nonmarine unit. All four formations were part of Beal's original single Rosario Formation. The total thickness of the Rosario Formation near El Rosario is 2900 m.

"[Fossil] collecting in the vicinity of El Rosario [has been] a continuing project since 1965...". Morris (1971) has described vertebrate fossils from the El Gallo Formation and says, "the fauna is small due to difficulties in collecting, scarcity of specimens, and the almost unique sedimentary environmental framework. Dinosaurian archosaurs related to *Lambeosaurus* of the Canadian Rocky Mountain foothills and Alberta plain are the most common herbivores. Some were 17 m. long and were more aquatic than terrestrial. A large carnivorous dinosaur has been recognized by isolated teeth as well as cranial material. It was morphologically near *Gorgosaurus* [The tyrannosaur *Albertosaurus S.F. tyrannosaurinae* is more widely recognized by its synonym *Gorgosaurus*.].

THE "TERRIBLE LIZARDS" DINOSAURS: Dinosaurs evolved from reptilian ancestors sometime in the Anisian stage of the early Triassic period about 235 million years ago and became extinct at the end of the Mesozoic during the Maastrichtian stage of the late Cretaceous 65 million years ago. The earliest dinosaurs were not really lizards at all but were members of a group of early crocodile-like thecodonts known as archosaurs. Eventually dinosaurs populated all of the large land masses except Antarctica.

Paleontologists have delineated two orders of dinosaurs — the Saurischia (lizard hipped) and the Ornithischia (bird hipped). It has been determined that if a strictly phylogenetic classification were used, it would be correct to say that modern birds are descended from the Saurischians and are living dinosaurs.

In 1822 the first fragmentary dinosaur fossils were found in England by the wife of Gideon Mantell, an English medical doctor and avid fossil collector. He named the dinosaur that his wife found *Iguanodon*. The discovery was first briefly described in an 1822 publication followed by a formal scientific publication of the discovery in 1825. The first dinosaur fossil discovered in North America was a hadrosaur (duckbill) tooth found in Montana in 1854 in the Judith River Formation by paleontologist Ferdinand Hayden. The first complete dinosaur skeleton, which was also a duckbill, was found in Haddon Field, New Jersey, in 1858 by William Parker Foulke. Until this time most dinosaur fossils which had been found were bone and tooth fragments and most were found in Europe.

Today, however, the American West presents the richest and most varied collection of dinosaur fossils anywhere in the world. The United States has become the center for the study of dinosaurs, and today most of the field research and the interpretation of the bones takes place in The United States. Near El Rosario, Baja California, several dinosaur bones and tooth fragments have been found and assigned to the Family Hadrosauridae.

Paleontologists attempt to recreate the history of past life from fossils. Some of the most exciting and challenging material they work with is related to the fossilized traces of dinosaurs. Fossil materials collected in Baja have contributed significantly to the increase of knowledge about the history of the geology, climate, and biology of the peninsula. Additional studies are needed to help clarify Baja's past. Do not collect Baja's fossil materials, since they are needed to illuminate Baja's past. It is against Mexican federal laws to collect the peninsula's resources!

MAMMALS: Although Cretaceous mammalian fossils are rare, they have been found in Baja. Specimens consist of isolated teeth but several jaws have been collected and are related to forms found in Cretaceous deposits which outcrop along the eastern side of the Rocky Mountains.

REPTILES: A crocodilian which is close in ancestry to the lineage which leads to modern alligators has been uncovered.

BIRDS: A very significant avian fossil has been collected. This specimen represents the only terrestrial bird recovered from Mesozoic strata other than the famous *Archaeopteryx* from the Solenhofen Limestone of Germany. Studies indicates that it will serve as a phylogenetic link between the Jurassic *Archaeopteryx* and modern terrestrial aves."

54.5 The canyon on your left at approximately 8 o'clock contains exposures of the Rosario Formation. In the right fork of this canyon fossils of Mesozoic *Ammonites* have been found high on the slopes and ridges on the right side of the canyon.

55.5 El Rosario is a small fishing and agricultural community. In the past prior to the completion of the paved highway in 1974 this small farming community was considered the last outpost of civilization before the trip into the remote "interior" of the peninsula. It was the last town until Rosarita, which was two hard days of travel down the unpaved old road. The road to the right at the corner leads to the ocean at the mouth of the Rio El Rosario. A branch crosses the arroyo to the old mission ruins and ends at Punta Baja.

56 The highway follows Arroyo El Rosario which is located on a recent river terrace, to Kilometer 61. Rocks of the Rosario Formation are exposed on both sides of the arroyo.

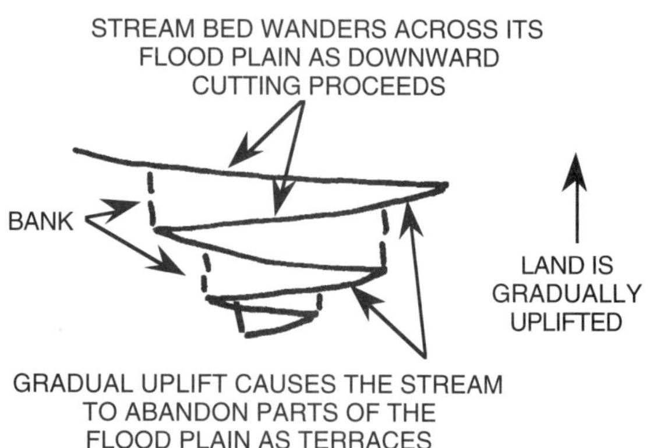

CROSS-SECTION OF EL ROSARIO VALLEY

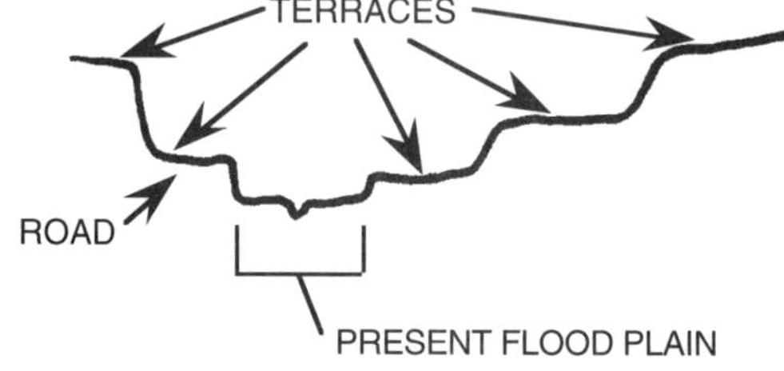

Across the Arroyo to the south there is a series of flat "benches". These benches are called river terraces and were formed by the downward cutting action of Rio del Rosario. The land in this area is gradually rising relative to sea level. This enables the streams to cut deeper into their beds which leaves behind parts of their former channel. The terrace which the highway crosses is one of these river terraces.

To the east of El Rosario along the left side of the highway farmers often cover the hills with chili peppers and allow them to dry in the sun before marketing them.

"CHILIES PEQUENOS" = SPICY, HOT CHILI PEPPERS!: Capsicum (chili) peppers, members of the potato family (F. Solanaceae), are used as a spice to flavor foods in both the New and Old World. There is a wide range of varieties. Peppers are New World natives that originated in Central Mexico and South America. They have been eaten for at least 7,000 years and have probably been cultivated for almost as long. Capsicum peppers are known from prehistoric burial sites in Peru, and they were widely cultivated before the arrival of Columbus in the 1490's.

The flavor and aroma of spices are due to essential oils. These oils are organic substances of varied composition which tend to have relatively small molecules. This makes them volatile (readily vaporizable at low temperatures). These oils belong to a group of hydrocarbons known as terpenes. The terpene of capsicum peppers is a volatile phenolic compound called capsaicin ($C_{18}H_{23}NO_3$), a substance so powerfully affecting human taste buds that even a dilution of 1:1,000,000 is detectable.

The greatest concentration of capsaicin is found in the pepper's placenta, the tissue which joins the seeds of the capsicum pepper to the fruit wall. Seeds of capsaicin peppers also have a lot of capsaicin. The fruit wall (ovary) has the least capsaicin. Because its presence or absence is due to variation in a single gene some peppers lack capsaicin. Many people who are not used to eating spicy peppers such as those encountered in Baja remove the seeds and placenta before eating or using capsicum peppers as a spice. This produces a milder flavor.

Capsicum peppers vary in size from large bell peppers, which are not peppery at all, to small extremely hot ("picante") red chili peppers known as "chilies pequenos" (*Capsicum annuum*). There are four closely related species involved in the production of economically useful peppers worldwide, and bell peppers and chili peppers may both be produced by the same species (*Capsicum annuum* or *C. frutescens*). In addition to their use as spices, capsicum peppers are rich sources of vitamin C .

58 South of El Rosario the vegetation changes dramatically with the appearance of the first endemic Cirios (*Idria columnaris*).

PENINSULAR ENDEMISM: Peninsulas such as Baja often have endemic (indigenous or native) plants and animals which are confined naturally to the peninsula and are not found anywhere else in the world. Due to the semi-isolation Baja California has experienced in past geological times, endemism is common. As a result many spiders, plants, and reptiles are peculiar to the Baja Peninsula. Endemism is especially evident among the cacti. Over 110 species cactus have been reported in Baja; 80 of these are found in no other parts of the world. Other notable Baja endemics which are commonly seen include the Cardon cactus and the Elephant Tree. This endemism contributes to the picturesque uniqueness of Baja's vegetation.

DESERT ANNUALS: After rainy periods, whose times and amounts vary from desert to desert, the landscape is covered briefly by colorful "blooms" of annual ephemeral desert wildflowers. Their showy variously colored flowers (primarily white and yellow) stand out vividly against the earth tones of Baja's desert scape. Wildflower blooms are most often seen in the early spring in northern Baja and in the summer in the southern part of the peninsula. Annuals cannot withstand drought and may lie dormant for decades until the right conditions occur for growth. Then the desert "blooms" with colorful floral displays. It is the function of the seeds to be the source for the next generation. These plants, which are so characteristic of the desert, evade the harshest desert conditions of Baja.

62 A high bridge crosses Arroyo El Rosario. The hurricane of 1967 did extensive damage to the El Rosario area. In early 1979 before the bridge was built, the paved highway was completely washed out, and some travelers waited for four days to cross. Vehicles either forded or were towed across the wash by a caterpillar tractor.

The dirt road up the north side of Arroyo Rosario leads to the small Rancho Provenir. The majority of the hills along Arroyo Rosario consist of the mudstones and sandstones of the Rosario Formation. For the next 20 kilometers the highway climbs through subdued exposures of the Rosario Formation.

EL CASTILLO

FAULT BEHIND CASTLE PARALLELS RIVER

SW NE

MUDSTONE HILLS BEHIND CASTLE

MUDSTONE HILLS BEHIND CASTLE

CONGLOMERATE AND SANDSTONE COMPRISE CASTLE

FAULT BEHIND CASTLE PARALLELS RIVER

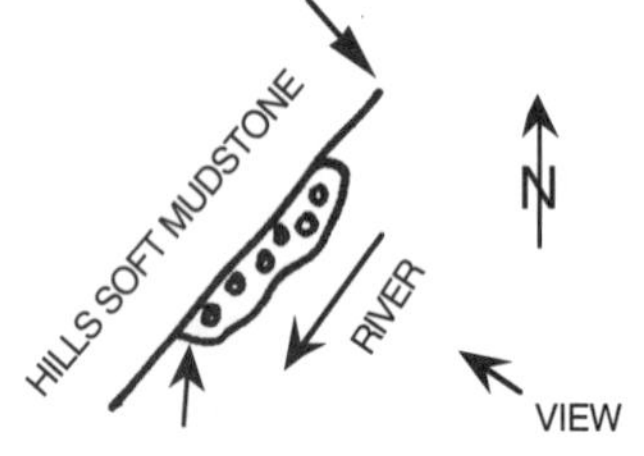

SLIVER OF RESISTANT CONGLOMERATES AND SANDSTONE FORMS CASTLE

63 **EL CASTILLO:** An interesting castle-like formation complete with battlements, which has been dubbed "El Castillo", can be seen upstream to the left as the highway crosses the bridge. Movement on a fault parallel to the arroyo has resulted in the veneering of the canyon side with resistant flat-lying conglomerates that form the castle. The yellow-brown band high on the hills is the boundary between the Cretaceous and Paleocene rocks in this area.

68 The roadcuts located on either side of this kilometer provide a good opportunity to stop to inspect excellent exposures of the mudstones of the Rosario Formation. Kilometer 68.3 has a good turnout for safe parking.

69 The sandstones and mudstones of the Rosario Formation are exposed in the roadcuts along this grade. The highway ascends a canyon and passes the first occurrences of specimens of Cirios (*Idria columnaris*) on the left.

71 A "grove" of cirios is growing on the hills to the left of the highway. Parry Buckeye, (*Aesculus parryi*) are growing on the northern slopes along this stretch of the highway

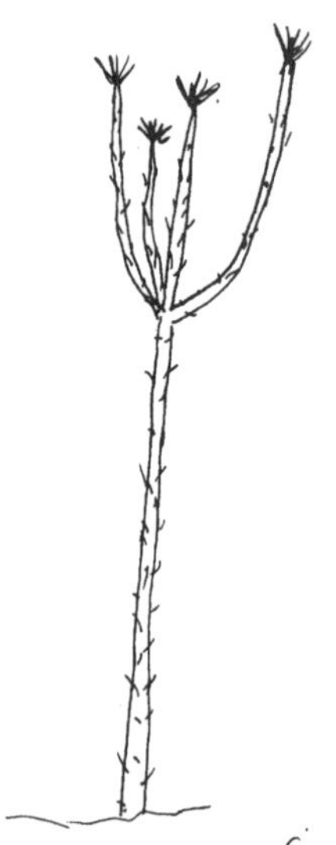

BAJA'S WEIRD CIRIOS: The cirio trees of Baja are claimed by some as one of the endemics of the peninsula. However, a small colony of cirios are also found on the mainland of Mexico south of La Libertad. It is one of the most distinctive and interesting plant species in Baja.

This tall candle-like plant derives its name "Cirio" from the Spanish word for wax taper (candle). It has also been called a "boojum" because of its similarity to the fanciful creature described in Lewis Carroll's "The Hunting of the Snark". Cirios have tall trunks which taper to a point approximately 20 meters above the ground in the largest specimens. However, due to the arid conditions of the peninsula, specimens rarely reach their maximum height. The whole plant looks somewhat like an upside-down carrot, with whip or rootlike limbs which curve up and down in a grotesque fashion.

CIRIO WITH JEEP

A story of doubtful authenticity has been told by some "Baja Travelers" in an attempt to explain the strange shapes of the whiplike grotesquely curving limbs.

This desert is very dry, so dry that when there is a fog (Neblina), there is more moisture in the air than in the ground. In order to survive the Cirio trees turn upside-down so that their "roots" are in the air in order to reach the fog. Since the fog layer is thin, the "roots" quickly grow above the fog and bend toward the ground to get back into the moisture-laden fog. As they grow toward the earth and grow below the fog they again turn upward away from the ground.

Because of its strange appearance, the cirio has often been called the most unusual plant of Baja. Cirios are related to the Ocotillo but have white instead of red flowers. During the dry seasons the small leaves of the Cirio fall, and the trunk is dormant and protected from desiccation by a thick waxy epidermis. When it rains, new leaves form from protected lateral buds. The cirio prefers to grow on the west-facing slopes of hillsides and alluvial plains from south of El Rosario across the peninsula through the volcanic fields and into the Gulf Coast Desert.

71.5 Good exposures of Rosario Formation conglomerates are visible in these roadcuts.

74.5 The highway crosses a large arroyo. There is a small turnout on the grade which provides a good view of the varieties of vegetation which are characteristic of the Vizcaino Desert Region.

DESERT PHYTOGEOGRAPHIC REGION: The Chaparral flora of the Californian Phytogeographic Region declines near Valle Santo Tomas. In the intervening 160 kms. between Valle Santo Tomas and El Rosario a transitional (ecotonal) flora which consists of a mixture of Chaparral and desert species predominates. Extending to the south from El Rosario east to the Gulf and to the south to La Paz is the drier Desert Region.

In contrast to the 25 to 50 centimeters of annual precipitation received by the more northerly Californian Phytogeographic Region, almost no rain falls in the Desert Region for two or more years. This region is truly a desert and is characterized by low humidity, widely fluctuating high ambient air temperatures, high surface and soil temperature, low organic content of the soil, strong winds, high mineral salts content, erosion by wind and water, poor drainage, and scarcity of water.

As the highway proceeds to the south through the Desert Region, it passes through three of the four subregions which are recognized and characterized by plant species

These three subregions are the Vizcaino Desert, the Gulf Coast Desert, and the Magdalena Plains Desert.

The highway between the Pacific coastal towns of El Rosario and Santa Rosalia on the Gulf coast passes through both the heart of the Vizcaino Desert and a portion of the northernmost extension of the Gulf Coast Desert.

The extremely dry **Vizcaino Desert** covers the vast plain in west-central Baja California and extends from El Rosario to the date palm oasis town of San Ignacio.

The floral dominants of the Vizcaino Desert Region are Cardon, Cirio, Century plants, Maguay, Datilillo, Agave, Yucca, Burbrush, and Ball Moss.

VEGETATION OF THE VIZCAINO DESERT

CIRIO

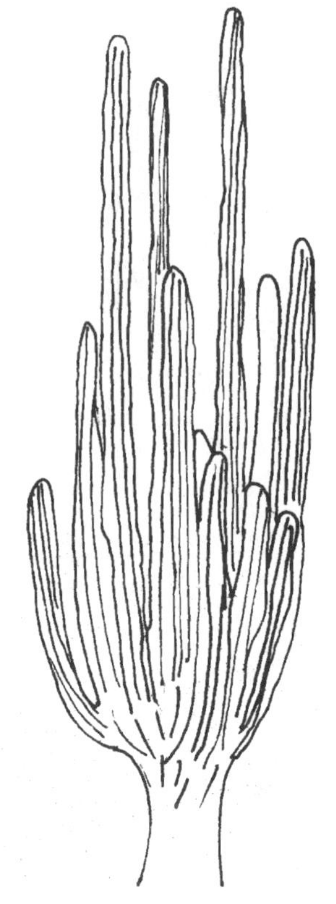

The **Cirio** or Boojum is the tallest plant of the visible flora. It looks like the tap root of a plant that has been turned upside down. *See* 3:69 to review the details about this strange plant. Cardon is a cactus similar to the Saguaro cactus of the Sonora Desert of mainland Mexico and the deserts of the southwestern U.S.

The Cardon (*Pachycereus sp.*) is a true Baja endemic. It is found from El Rosario to the tip of the peninsula and is the most widely spread of the peninsula's larger vascular plants. Since the cactus lacks leaves, photosynthesis takes place in the modified epidermal cells (chlorenchyma) of the trunk. The trunk is a true cladophyll (stem acting like a leaf). Stands of Cardon are called Cardonals

Nests of hawks and Osprey are often seen in the upper branches of this giant columnar cactus. Hawks, buzzards, and ravens use the Cardon as hunting, resting, or sleeping perches.

CORDON

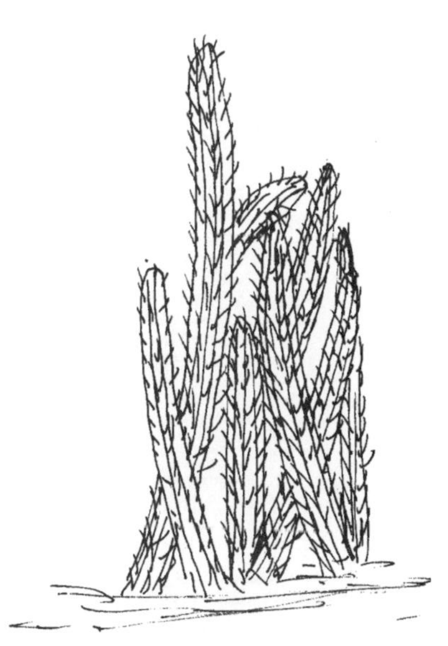

PITAYA AGRIA

Pitaya Agria is a dark green-gray stemmed cactus which grows in dense thickets from Ensenada to the Cape Region. The spines are reddish-gray with darker tips. The fleshy red fruit is edible.

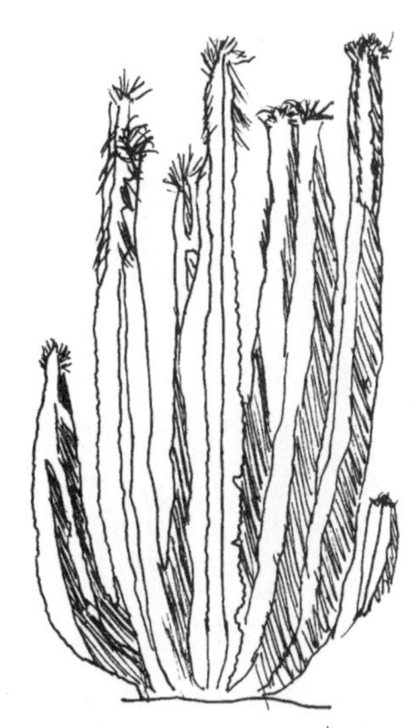

GARAMBULLO

"Old Man" Cactus or Garambullo is an erect colonial cactus "tree" which grows to 4 meters tall and usually branches near the base. The tips of the star-shaped stems are densely covered with many long gray coarse hairlike spines which give the cactus a whiskered "old man" look. The fleshy red fruit is edible but is not nearly as good as the fruit of Pitaya Agria. It occurs from south of Rio Del Rosario to the Cape Region.

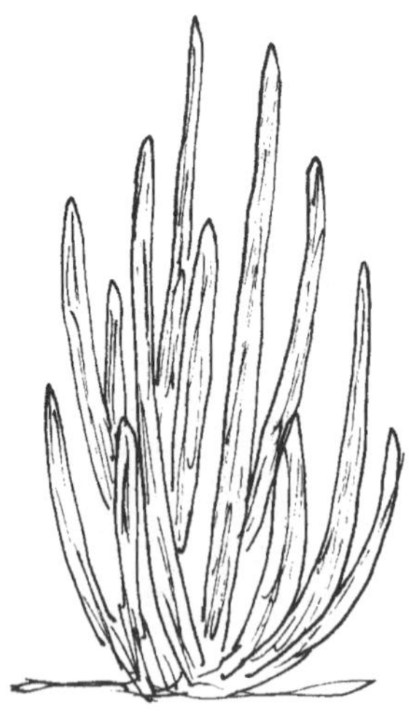

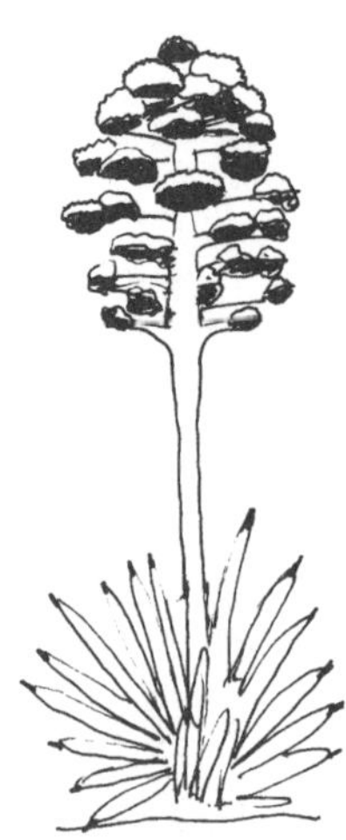

ORGAN PIPE CACTUS
PITAYA AGRIA

Organ Pipe Cactus is a many branched erect cactus without a main trunk. It might be confused with the cardon; however, the Organ Pipe Cactus, which branches nearer the ground, lacks the main trunk characteristic of the Cardon. It looks like a set of organ pipes. The fleshy watermelon flavored red fruits are used as a food source in the late summer and fall.

AGAVE

Agave or Century Plants are members of the Amaryllis family. In Baja four species of the genus *Agave* are commonly referred to as Agave or century plants. It takes determination and a brief comparative study to tell the species apart.

The various species of *Agave* have a long history of usefulness in Mexico. In the Yucatan Peninsula of mainland Mexico cultivated species of *Agave*, also known as "green gold", were once extensively cultivated for the fiber henequen (sisal). Botanically the *Agave* fibers are long sclerenchyma, modified parenchyma fibers, like the "strings" in celery. Until the end of WW II the Yucatan Peninsula supplied much of the world's henequen (sisal) fibers and which were used in the production of ropes, rugs, twine, and wall hangings. The fermented products from the plant are still used to produce the alcoholic beverages of Tequila (distilled in the Tequila region of Jalisco, Mexico), Pulque (brewed throughout much of Central Mexico), and Mezcal (distilled outside of the Tequila region of Jalisco, Mexico).

The Agave is utilized for food. The immature flowering stalks and basal crowns are roasted. The seeds were ground into a meal.

BALL MOSS

Ball Moss or Gallitos, a member of the pineapple family, is an herbaceous epiphyte that commonly grows on other plants but does not damage it or derive nutrition from it. In this region it is a common commensalistic epiphytic plant which grows on Cirios, cacti, shrubs and even telephone wires (*See* 16:16.5). Some travelers mistakenly identify Gallitos (little chickens) as birds nests. Ball Moss is disseminated from plant to plant by wind-borne sticky seeds or by birds' feet or bills.

Ramalina is a lichen. Lichens are a partnership formed by a combination of two plants which grow in such close companionship that their separate tissues can only be determined under a microscope. This partnership consists of microscopic photosynthetic green algal cells which live inside the cells of a non-photosynthetic fungus (mushroom). The relationship formed is known to botanists as a mutualistic symbiosis. A lichen is an algae and a fungus that have "taken a liken to each other"; both plants benefit. The fungus provides water and protection for the algal cells. In return it utilizes the sugars, which are produced by the photosynthetic algae, for its source of energy. When times are difficult, the fungus will digest the algae. Instead of being "eaten out of house and home", the algae is eaten by its home!

In this area of Baja lichens grow epiphytically on Cirios or encrust boulders. Different species of lichen are visible as vivid red, silver-gray, and even black splashes of color.

76 Mesa la Sepultura "tomb" is the flat feature which is straight ahead. This mesa is the type area for the Sepultura Formation, named by Santillan and Barrera in 1930. The highway continues to pass through the Rosario Formation which, at this point, extends far inland (northeast).

77.3 This side route leads across Mesa San Carlos into Arroyo San Fernando to Abelardo Rodriguez (31 kilometers) and San Vicente. It ends on the coast at the base of Mesa San Carlos at Puerto San Carlos (60 Kilometers). It is a long dead-end route, but the solitude and surfing make the trip worthwhile.

Highway curves to the northeast and drops through prominent exposures of conglomerate into the main wash of Canada El Aguajito near the old Rancho of El Aguajito. Conglomerates which contain lenticular sandstones of the Rosario Formation are exposed in many places in this wash.

As the highway drops into the wash, Cardons, Rabbit brush, Burbush, Burrow Weed, and Wash Woodland-type vegetation predominate the landscape. On the left, just after crossing the wash, the typical Vizcaino Desert flora is represented by Cardon, Cirio, Pitaya Agria, Ocotillo, Garambullo, Agave, Organ Pipes, Cheese Bush, Burro Bush, Broom Baccharis, Mesquite, and epiphytic ball moss which grows on the Cirios. If you take a walk through the vegetation, you may see and hear the flitting red-capped Gila Woodpecker (*Melanerpes uropygialis*).

The **Gila Woodpecker or Carpintero de Gila** is a medium-sized, red-capped woodpecker with a black and white barred back. It is often seen perching or flitting around among willows and cottonwoods in riparian habitats and in low desert cardonals and stands of mesquite in Baja. This is the only desert woodpecker with a plain gray-brown belly, head, and neck and white wing patches which show during flight. They frequently bore nest holes in Cardons, mesquites, and cottonwoods throughout the peninsula. Gilas feed on insects and the fruits of mistletoe and cacti.

80.2 The contact between the Rosario and the Alisitos Formations is two thirds of the way up the grade, east of the wash.

82.2 At the top of the grade is a sweeping view. A series of concordant summits of the mesa tops can be seen to the northwest stretching toward the beach at El Rosario. To the north and east is the granitic spine of Sierra San Pedro Martir. Picacho del Diablo, 10,126 feet elevation, is the highest peak in Baja California. To its right is Pico Matomi, a Miocene volcanic neck of andesite porphyry. To the east is the Sierra San Miguel. To the south is Mesa la Sepultura.

84 Along this ridge to the right are beautiful views of Mesa la Sepultura. Mesa la Sepultura and Mesa San Carlos are composed of the Cretaceous marine Rosario Formation in the lower parts and are overlain by the Paleocene-Eocene marine Sepultura Formation (Santillan & Barrera, 1930). The Sepultura Formation consists of conglomerates which grade into mudstones with conglomerate channels to the west. This Cenozoic section extends inland 30 kilometers. It becomes thinner and more coarsely clastic and nonmarine with sandstone, conglomerate and scattered thin shale partings which pinch against an irregular buried topography.

Due to runoff, exposure, and winds, life on this ridgeline is extremely harsh. The vegetational "cover" is sparsely represented by Cirio, Cardon, and Agave, all true xerophytes.

Xerophytes are plants. In Baja xerophytes include succulents, such as various cacti or members of other families which have fleshy stems and leaves which enable them to store water for a long time. They frequently have shallow root systems and are able to utilize the soil moisture from light rainfall, heavy dew, or fog. Such plants take advantage of the little precipitation which falls in the desert areas of Baja by storing precipitation, dew, and/or fog in their pith and cortex parenchyma tissues for months or even years. Many succulents such as cacti have leafless and ribbed stems, two more adaptations for survival in desert environments. Leaflessness reduces the surface area through which water is lost by transpiration. The ribbed stem allows the stems to swell like an accordion and store water when it becomes available.

85.3 The metavolcanic rocks of the Alisitos Fm. are well exposed on the right side of the highway. To the right at 2 o'clock are the white buildings and scar of the La Turquesa turquois mine.

LA TURQUESA: Turquoise, a light-blue to blue-green triclinic phosphate mineral, is formed as a secondary deposit which fills cracks and shears in the metavolcanic rocks of the Alisitos Formation. Turquoise, which is the birthstone for December, usually occurs in cracks and as reniform masses with a botryoidal (the form of a bunch of grapes) surface in the zone of alteration of aluminum-rich igneous rocks.

The La Turquesa mine became a series of "gopher (go for) holes" as the miners followed the illusive mineral. They mined only what was absolutely necessary. Several other abandoned mines are located in the area.

86 Exposures of reddish soils of the Alisitos Formation can be seen along the sides of the highway.

86.4 There is a rather interesting little gorge in the Alisitos Formation to the right.

There is an obvious change in the vegetation from the mudstones and conglomerates of the Rosario Formation to the reddish-brown soils of the Alisitos Formation. This change is due to differences in soil chemistries, soil moisture, soil salinity, and other soil edaphic factors.

The vegetational cover consists of Teddy Bear Cholla, Barrel Cactus, Cirio, and variously colored crustose lichens which look like paint that has been "splattered" randomly on the rocks. Each color is a different lichen.

88.6 The side road is the turnoff which leads to the turquois mine, La Turquesa. The highway descends the new Aguajito grade through adamellite and the metavolcanic rocks of the Alisitos Formation.

90 To the south is the bedded Alisitos Formation which dips slightly to the south. Excellent exposures of the Alisitos Formation, which is composed of rhyolite, ignimbrite, basalt and andesite, outcrop in this area.

Lemonadeberry (*Rhus sp.*) is the large darker green plant which grows alongside the highway. Indians prepared a tasty lemonade-like drink by soaking the bright red berries of this large shrub in water. Because the berries contain a large amount of malic acid, they act as a diuretic (increasing the flow of urine).

91 The zigzag pattern of the old road can be seen to the right. Fierro Blanco in his book *The Journey of the Flame* writes of the last moments of Señor Don Juan Obrigón. The Flame, who died alone at the Great Cardon, near El Rosario, with his face turned to the south. Could this be the "Great Cardon"?

LARGE CARDON AND CAROL IN 1967

The Cirio on the right contains a raptor nest which has been used by the Red-Tailed Hawks which are commonly seen in this area.

92 The old road crossed a flat valley on a Tertiary conglomerate which may have been an ancient river channel. This conglomerate may be correlative with Cretaceous age conglomerates which formed when the area was first elevated and eroded due to the pushing of the Pacific plate under Baja California.

95 One of the roads to Cerro Blanco is to the left down this arroyo. There are several old mines, including a copper mine, in the area.

Beavertail Cactus, Agave, Barrel Cactus, Pitaya Agria, Broom Baccharis, Toyon, sparse Cardons, and tall Cirios which are covered by the lichen *Ramalina*, comprise the vegetational cover of this portion of the Vizcaino Desert.

97 The highway crosses a prominently jointed adamellite (looks like a dike). The old unpaved road also crossed at this very rough spot.

101 This kilometer marks the beginning of a large Cardonal and Cirio forest. The vegetation of the area is typical of the Vizcaino Desert flora and is dominated by Desert Thorn, Mimosa, Mesquite, Desert Mallow (along the pavements edge), Cirio, Datilillo, Garambullo, Beavertail, Jumping Cholla, Barrel Cactus, Tamarisk (in washes), Yucca, Cardon, and Palo Estribo.

The highway passes Rancho Arenoso, one of the numerous, old-time cattle ranchos of the peninsula. The old road led right to the front door of the ranch house where earlier Baja travelers stopped for cold refrescos and a respite from the rough bone-jarring road.

106 Slightly past this kilometer sign, approximately two o'clock to the right, there are a couple of low hills. The large four-branched Cardon visible on top of the higher hill makes a scenic photograph in the right light. In the spring these attractive Cardon produce numerous beautiful flowers and globose fruits. Look for woodpecker holes in the cactus.

Along the highway there are large barrel cactus which are known as the "compass plants" of Baja.

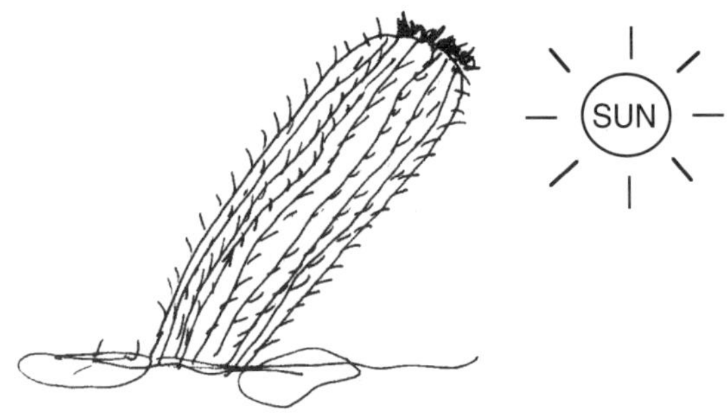

Because of their tendency to grow toward the light of the sun to the southwest (to reduce the harmful effects of the sun during the hottest times of the year), they are called the "compass plant".

BARREL CACTUS

The tops of the cactus are surrounded by reddish-thorns. In the late spring they have a ring of beautiful waxy yellow flowers. In an emergency the Barrel Cactus will yield approximately a pint of alkaline juice from its pithy cortical parenchyma (water storage tissues of the stem).

108 The highway roughly follows a strike valley (Strike = horizontal compass bearing of the beds) in the metavolcanic rocks of the Alisitos Formation. In the hills ahead and to the right two interesting erosional curves in the dipping limestone beds are visible. The beds are not folded.

DIP SLOPES WITH EROSIONAL CURVES

112 The highway enters a strike valley in the Alisitos Formation with beds which dip toward the east at approximately 45 degrees.

Prominent ridges which contain limestone are exposed on both the right and left sides of the highway. The highway roughly parallels the strike of the Alisitos Formation for several kilometers. Look to the left at the long strike ridges (45 degree dip) across the valley in the near distance. This is reminiscent of the plateau area of northern Arizona and southern Utah, except the age of the rocks and the vegetational cover differ.

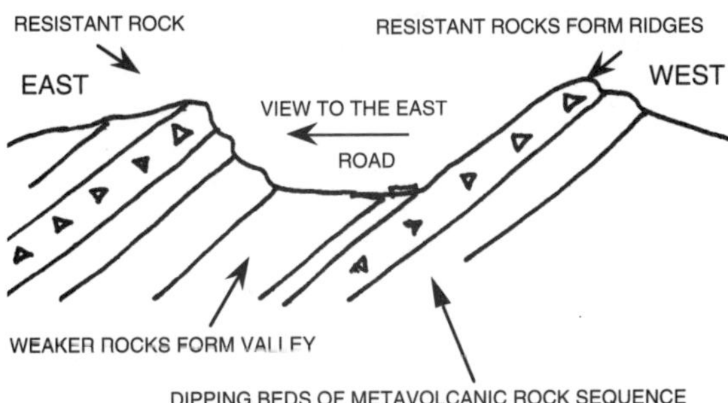

Metavolcanic rocks: Large areas of Metavolcanic rocks without fossil or radiometric dates are exposed throughout much of the Pacific slope of northern Baja California. South of the Agua Blanca Fault, the volcanic sequence includes a variety of sedimentary strata. Limestone, calcareous siltstone and mudstone are interbedded with volcanic sandstone, volcanic conglomerate, tuff and volcanic breccia, and represent a wide variety of depositional environments. Deposits range from deep to shallow marine and nonmarine, from coarse sedimentary breccia to limestone, and from basalt to rhyolite. Andesite and andesite breccia are the predominant volcanic rocks. The entire sequence is thought to be tens of thousands of meters thick. It varies widely in stratigraphic components from area to area. No two measured stratigraphic sections are alike. Mapping and stratigraphic work will be required before subunits can be recognized and correlated (Gastil *et al., 1975).*

113 Yuccas begin to be more abundant.

YUCCAS are 1 to 5 meter high shrubs of the lily family with sharply pointed leaves and a large flowering stalk which support thickly clustered, terminal panicles of whitish flowers. The flowers are pollinated by tiny night flying Pronuba moths. Neither the Yucca nor the moth can propagate its species without the aid of the other. The moth larvae feed exclusively on the Yucca seeds and the Yucca flowers are only pollinated by the moth. Eliminate one of the symbiotic mutualists and the other will die.

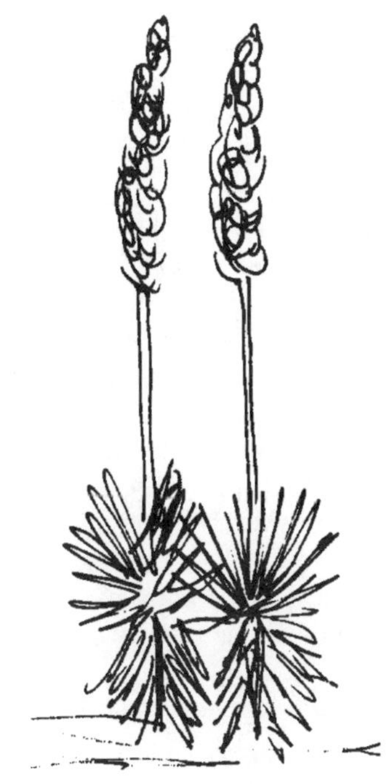

Both man and cattle make use of parts of the Yucca. The stalks are rich in sugar; the saponins stored in the roots are used as a substitute for soap. (Remember Yucca-Do-Shampoo?) Saponins are plant glucosides that form a soapy colloidal solution when mixed and agitated with water. Saponins are used in detergents, synthetic sex hormones, foaming agents, and emulsifiers. The flowers are edible. The green cucumber-like fruits can be eaten raw or roasted. The long sclerenchyma leaf fibers were used in making baskets, sandals, and mats.

114 Turnoff to Mision San Fernando de Velicata (5.8 kms.) down the arroyo west of El Progresso. It was founded in 1769 by Father Junipero Serra as he traveled north to Alta California. It is the only mission in Baja California established by the Franciscans. It has deteriorated into low walls and mounds since its abandonment in 1818.

116.5 This is the new Rancho El Progresso.

118 The highway crosses a very lush wash woodland of Mesquite

119 The highway ascends to the top of a small grade and passes through the metavolcanic rocks of the Alisitos Formation. The old road went to the right, around the hill, and up the wash while the paved highway goes to the left, over the rise, and across the hill which avoids the flood waters that run seasonally through the wash.

120 View to the north is of Valle El Renoso. The view to the south is the plain of Llanos de San Agustin.

A Pleistocene lake bed: The light-colored brownish-tan (buff) sandstone, siltstone and limestone which occupies the bottom of Valle Santa Cecilia were part of one of a series of Pleistocene lake beds that occurred in this part of the peninsula. Various freshwater vertebrate fossils that include fresh water turtles have been recovered from the sediments of the lake. The possibility for finding more vertebrate remains is fairly great in this area. The sediments of this "fossil" Pleistocene lake are fine-grained, and this part of the highway was very dusty when dry.

PLEISTOCENE LAKE BED SEDIMENTS

122 The highway traverses the flat upper surface of the sediments on what was the lake bot-tom. On the north side of the highway the shoreline of the lake is visible at the base of the low hills where it was in contact with the meta-volcanic rocks of the Alisitos Formation. In the distance, on the south side of the highway, are low rolling hills that are the fluvial part of the Paleocene-Eocene Sepultura Formation. This relationship continues for several kms. along the highway.

The vegetation cover of the lake sediments is primarily Creosote Bush (*Larea tridentata*).

CREOSOTE BUSH

CREOSOTE BUSH is a very common resinous desert shrub of the southwestern U.S.A. and Baja. Unlike many other desert plants, it is green the year around. The waxy coating on its leaves reduces water loss by transpiration and allows the shrub to withstand long periods of extreme drought. The strong-smelling resinous leaves resemble the odor of the coal tar distillate-creosote. Notice that very few plants grow under the Creosote Bush. The roots of this shrub produce a toxic substance that inhibits the growth of other plants. After heavy or frequent rains, when it is leached from the soil, it permits the growth of desert annuals. As the soil dries, the inhibitor accumulates and poisons the outsiders. This phenomenon is known as allelopathy, and it helps to eliminate the competition for water in a dry desert environment. Creosote Bush was considered a cure-all by Baja's early Indians, and early Spaniards used a decoction of this shrub to treat sick cattle and saddle-galled horses. In more recent times a leaf extract has been used to delay or prevent butter, oils, and fats from turning rancid. Creosote Bush leaf extract is not widely used, because the compound is now synthetically manufactured.

125 The flat plains of this region are part of an exhumed erosion surface.

An Exhumed Erosion Surface This area in the middle of the peninsula consists of flat plains fringed by "not-too-steep looking" hills. The Llanos de San Agustin were base-leveled; that is, they were eroded to a relatively flat surface by streams during the Early Tertiary. The remnants of the deposits from one of these streams can be seen as conglomerates on the road to Santa Catarina. This base-leveled surface was then covered by Miocene fluviatile sedimentary rocks, basalts, and rhyolites which stretched across the peninsula. The area was not uplifted until sometime during the last 10 m.y. As a result of this uplift erosional processes stripped the covering rocks and "exhumed" the Early Tertiary surface, leaving lava-capped mesas scattered over the area as erosional remnants. Later impounded drainage filled the low places with lakebed sediments which are being dissected by present streams.

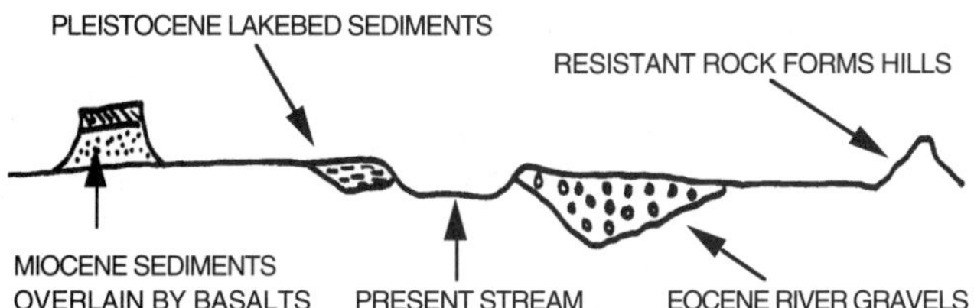

STYLIZED SECTION OF LLANOS DE SAN AGUSTIN

The vegetational cover is sparse and is represented by a few small Cirio, Cholla, Ocotillo, *Agave*, Creosote Bush, Yucca, and Mesquite.

Against the base of the metavolcanic hills the flora is dominated by bright green stands of Mimosa, which delineate the course of the wash which is visible from Kms 122 to 135.

126.1 The buildings along side the highway are part of the new Rancho Penjamo.

Several more of the old ranchos dot the old road on the left. The paved highway is approximately two kilometers south of the old road at this point. The Highway follows the flat surface of the Pleistocene fluvial lake beds discussed at Km 120.

127.8 The side road to the right leads southwest to Santa Catarina and Puerto Catarina on the coast. The Santa Catarina road follows the coast for many kilometers to the south and finally rejoins the highway near El Tomatal (4:68.5). It is a fairly rough, but quite picturesque, side road. It is only recommended for 4-wheel drive vehicles and group travel. We spent three weeks and four days camped on this road near Arroyo San Jose. Only two parties passed us during that time.

129 This is a good vantage point for the Llanos de San Agustin. The flat llanos were mantled by a blanket of fluviatile sedimentary rocks which were extensively covered by Miocene rhyolite flows. The little flat mesas and buttes that seem to stride the highway are remnants of these rhyolitic flows.

135 The flat reddish-colored mesa ahead is Mesa Redonda; it is capped by resistant volcanic rocks of Miocene age. Most of the flat hills in this area are mesas which are capped by these same resistant volcanic rocks. Another large raptor nest is visible in the large Cirio on the right side of the road at the base of Mesa Redonda.

137.3 The first Elephant Trees which are visible from the highway are growing on the slope to the right. These unique trees will be discussed at 4:9.

138 At the crest of this low pass there are exposures of gneisses and schists in the roadcut.

141 The windmill of the rancho of old San Agustin is visible about 1 kilometer to the left. It has a deep well with very good water. This well supplied water to the onyx mines at El Marmol.

142 As the highway crests over a rise at the top of the hill there is another good view of the flat plain of Llanos de San Agustin and the extensive flat Miocene volcanic mesas which are located on either side of the highway.

144 Turnoff to the abandoned El Marmol Onyx Quarry (15 km) at El Marmol (marble). The road to the mine is usually graded and should be passable with reasonable clearance passenger vehicles.

EL MARMOL

EL MARMOL ONYX SCHOOL HOUSE

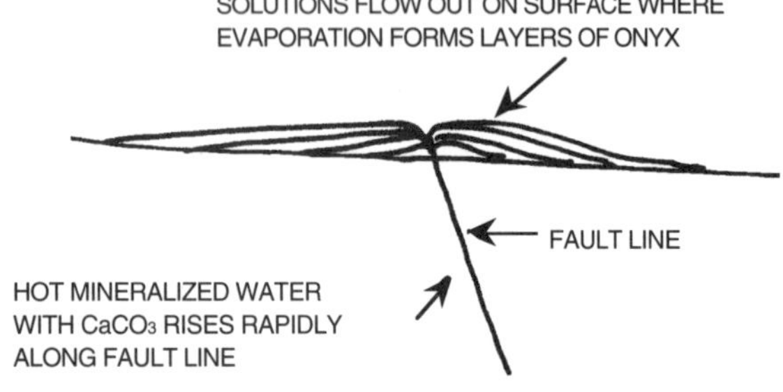

EL MARMOL ONYX DEPOSIT

EL MARMOL is the location of hot springs and a Travertine (onyx) deposit which was mined between 1900 and 1958 by the Southwest Onyx and Marble Company. Travertine is usually formed as a precipitate in a hot spring. Hot mineralized solutions from the El Volcan hot springs rise to the surface along a fault line, flow out on the surface, and evaporate as they cool.

Travertine is deposited in layers. There is a large amount of travertine at the abandoned mine. The old onyx school house is still standing and an interesting graveyard is located nearby.

EXHUMED EROSION SURFACE

145 The highway continues over Llanos de San Agustin. To the north (left rear) 40 kilometers in the distance are the two high peaks of the Sierra San Pedro Martir: the granitic pluton of Picacho del Diablo, 10,126 feet, the highest peak in Baja California, and Pico Matomi, a Miocene volcanic core of andesite porphyry, to its right. To the far right (west) are irregular hills of the meta-volcanic Alisitos Formation.

147 For years there has been a Red-Tail Hawk nest in the large Cardon just to the right of the highway. Careful observation of other large plants along the Highway may reveal other nesting sites of predatory birds (*See* 4:128 on Ospreys).

PREDATORY BIRDS OF BAJA: Over 270 species of birds of prey hunt in the daylight in the world. Over 140 species of owls are the world's nocturnal predators. In Baja there are 16 daylight and 4 nocturnal species of predators.

Predatory birds exhibit numerous anatomical and behavioral adaptations which enable them to lead a predatory life-style. For example, birds of prey generally have long curved talons for seizing their victims and strongly hooked beaks for tearing them apart. They are masters of soaring and swooping (diving). Baja's Peregrine Falcon (*Falco peregrinas*) may reach 175 mph as it dives for its prey. In certain hawks the eyes are larger and more acute than those of humans; like man's, they are binocular instead of being set on the side of the head as in most non-predatory birds. The most commonly seen predators along the Highway are Red-Tail Hawk (*see* 14:165), Red-shouldered Hawks (*see* 3:147), Northern Harriers (Marsh Hawks) (*see* 12:145), Ospreys (*see* 4:128), and Turkey Vulture (Buzzards) (*see* 8:36). Each will be discussed at kilometer marks where they have been commonly seen.

Red-shouldered Hawk or Gavilán ranero is distinguished by its reddish shoulders, narrow white bands on the wings and tail. Viewed from below, their underparts and wing linings are uniformly reddish. Their flight is accipiter-like, with several quick wingbeats and a glide. Red-shouldered hawks inhabit mixed woodlands and are often seen near streams. They hunt from a perch for snakes, frogs, mice, and young birds.

LAVA MESA ON EROSION SURFACE

148 The prominent flat-topped mesa to the right is Mesa las Palmillas; it is covered with beautiful stands of Elephant Trees and the weird Cirios. This is a good point to take a photo stop and view these two interesting trees.

152.5 As the highway passes through this roadcut, a well-developed Caliche layer can be seen. The mission padres used Caliche layers for lime to make mortar for the missions.

CALICHE refers to gravel, sand, or clay cemented by calcium carbonate. The surface of the soil is dry most of the time and there is a high percentage of calcium carbonate in the ground water. The dry soil acts as a wick to draw the water upward where it dries and precipitates the calcium carbonate as Caliche deposits near the surface of the soil in the B horizon.

SOIL HORIZONS

Pedocal soils generally have the following structures:

O-horizon-	Thin layer of fresh leaf drop and partially decomposed organic material.
A-horizon-	Decaying organic materials mixed with mineral material. This is a rather nutrient-void horizon because water which is pulled down by gravity carries most of the organically derived nutrients from the A-horizon and deposits them in the B-horizon below.
B-horizon-	Calcium and magnesium carbonate accumulate in this layer due to the lack of abundant water.
C-horizon-	Partially weathered bedrock.
Bedrock	Unweathered rock. (Barney and Fred)

PEDOCAL SOILS - In North America there are three major types of soils: laterites, pedalfers, and pedocals. Baja's soils are primarily pedocals. Pedocals form as calcium carbonate accumulates in the "B" horizon. These soils occur in areas where the temperature is generally high, the rainfall is less than 100 cm, and the upper level of the soil is dry most of the time (Baja!). Under these conditions there is not enough water remove the calcium and magnesium carbonate.

Because of the scant rainfall in Baja, the processes of chemical weathering occurs very slowly, and clay is produced less rapidly than in wetter environments. As a result, pedocals have lower percentages of clay than other soil types. The climate of Baja produces red and gray nutrient-poor desert soils, and the vegetation which grows on them is mainly grass, brush, and cactus. The plants that grow in pedocals often show the effects of poor soil.

153.5 Some of the first bouldery granitic rock outcrops, of the La Virgenes region are seen under the volcanic-capped mesas, visible to the right of the highway. This region, known as Las Virgenes, is full of large, spectacularly picturesque, granitic rock formations and many varieties of cactus and other desert vegetation. This region is also known as the Catavina Boulder Field.

154 Notice the large numbers of Barrel Cactus slanted toward the southwest in an effort to reduce the harmful effect of prolonged exposure to the southwestern suns rays during the hottest months.

155 At this kilometer there are granitic rocks with a knobby appearance that are exposed under volcanic rocks. The knobby granitic rocks are part of the Peninsular Range Batholith.

157 The vista opens to the south to reveal numerous lava-capped mesas with cinder cones on top of the picturesque bouldery outcrops of the Las Virgenes area.

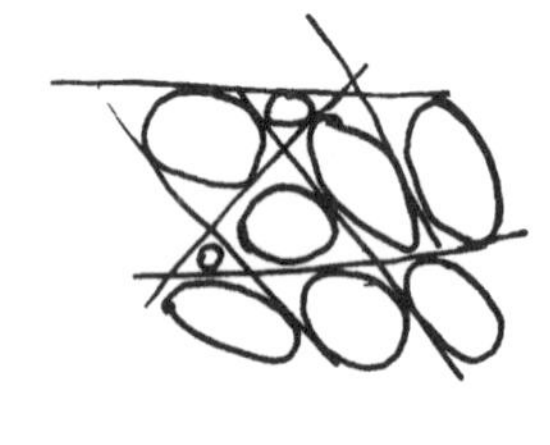

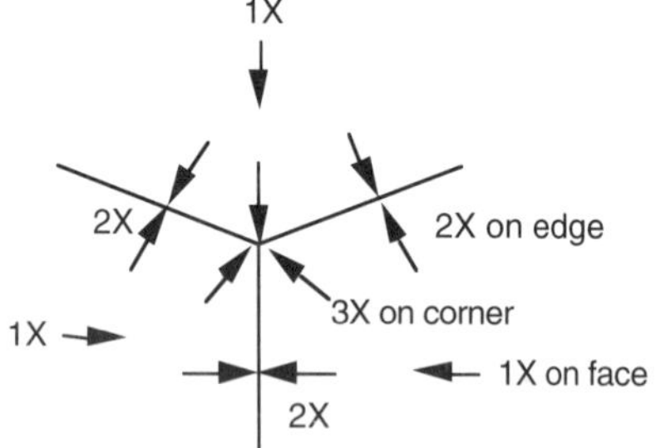

Spheroidal weathering can easily be explained. This area receives little rainfall; weathering proceeds slowly on the surface of the rocks. Since there are three surfaces at corners and two at edges, the rocks tend to weather into spheres. The weak running water carries the finer particles away which leaves the rounded boulders.

162 Baja travelers have named the small conical hill on the right (west) Mammalary Hill.

162.5 This side road goes to La Bocana and Rancho San Jose (95 Km). Get a guide and follow Arroyo La Bocana to the southwest for views of prehistoric painted Indian rock art (pictographs).

A display at the Pariador gives some information about the "Zona Arqueologica" in this region. Rock art can also be seen in Arroyo El Palmarito (Km 170).

Adopt the motto of naturalists everywhere: "take only pictures and don't even leave foot prints" so that others may continue to enjoy.

There is a view of the bouldery granitic tonalite terrain with basalt-capped hills of the Las Virgenes region. The peaks to the right are the Alisitos Formation. The metavolcanic rocks have yielded fossils from the Boca San Jose area.

AMMONITE FROM BOCA SAN JOSE

The vegetation is typical of the Vizcaino Desert flora and consists primarily of Cirio, Cardon, Elephant Trees, Garambullo, Cholla, Creosote Bush, Barrel Cactus, annual composites, Teddy Bear Cholla, an occasional Creosote Bush, Ephedra, *Agave*, Jojoba (goat nut), Pitaya Dulce, Pitaya Agria, and scattered Acacia, Datilillo, and Burro Weed. The orange "hairlike" plant which grows parasitically on the Elephant Trees is Dodder ("Witches Hair"). The opportunistic Brittle-Bush is on the highway edge. Smoke Trees and Mesquite are restricted to the washes; they look like smoke from a campfire. Smoke Trees are common plants of Wash Woodlands.

CORDON - CIRIO - MESQUITE

CHOLLA

There is evidence of mammals feeding on cactus in this area. This feeding on cactus may be to obtain metabolic water as well as food. Most Cardons bear scars from woodrats below 5 to 6 meters. Skeletons of Barrel Cacti are commonly hollowed out by small rodents and jackrabbits. Over half of the Chollas are actively fed upon by jackrabbits. The Old Man Cactus is fed upon by several animals.

The vegetation is more dense in the Catavina area. The boulders shed rain to provide more water for plants rooted near the boulder, thus increasing the net rainfall.

HEDGEHOG CACTUS

ARROYOS are gullies cut by intermittent streams and flash floods. Baja's arroyos, which are normally dry, are infrequently filled with water after it has rained for a few days. At these times there are flash floods which fill the arroyos from wall to wall with violently moving, sand-laden waters. The washes may run for one to several days depending on the amount and duration of the preceding rains. Just a few days after these flash floods, the arroyos will again be as dry as if they had never conducted a single drop of water.

165 Around the next curve the vista again opens with dark basalt which covers the high hills in the background and pinkish granites with spheroidal weathering in the foreground.

166 Some particularly picturesque rock formations and flora are found along this stretch of the highway. Good camping is available anywhere along the highway in this area especially along part of the old road which winds close to the highway. (Keep a clean camp and haul out all trash.)

This area is considered by many travelers to be one of the most scenic in Baja because of its bouldery rugged beauty and the arid starkness of the Vizcaino Desert and its unique vegetation.

ELEPHANT SEAL ROCK AND CREOSOTE

Birds from the California Phytogeographic Region are often seen here due to the proximity of that region to the north (*See* Phytogeographic map) as well as the higher moisture produced by the rocks. A winter biology study by the College of Idaho produced the following bird list in the Catavina area. They felt that more birds were seen in rocky sites because there is more protection from wind, more diversity of perches, and more abundance and variety of plants.

BIRDS OF THE CATAVINA REGION

* Anna's Hummingbird
Ash-throated Flycatcher
Black Phoebe
* Black-chinned Hummingbird
Black-tailed Gnatcatcher
Black-throated Sparrow
Blue-gray Gnatcatcher
Brown Towhee
Cactus Wren
California Quail
Cassin's Kingbird
Common Raven
* Costa's Hummingbird
Gambel's Quail
* Gila Woodpecker
Gilded Flicker
Ladder-backed Woodpecker
Mockingbird
Phainopepla
* Red-shafted Flicker
* Roadrunner
w Robin
Rock Wren
Rod-tailed Hawk
Say's Phoebe
w Scott's Oriole
Sparrow Hawk
Turkey Vulture
Verdin
w White-crowned Sparrow
White-throated Swift

* Displayed courtship or nesting activity
w - indicates possible wintering species

---

170 The highway begins a descent into Arroyo Catavina (Arroyo El Palmarito). To the left, flat-lying volcanic rocks cap the erosion surface of the granitic rock.

Take some time in this area to explore and take pictures. In this region dawn, dusk, and afternoons are particularly picturesque and photographic times of the day.

SPHEROIDAL GRANITICS & BLUE PALMS

171 Along the left side of the highway, a short distance up Arroyo Catavina a large lush stand of native palms are growing; they include specimens of fan palms (*Washingtonian falifera*, and *W.*

robusta), and blue fan palm (*Erythea brandegeei*). The presence of palms in the desert often indicates the presence of perennial water.

Archaeological evidence indicates large numbers of prehistoric native Indians once occupied the area and made use of the perennial water located in this region. In boulder areas west and north of Pariador Catavina, stone chips can be seen littering the surface of the ground. Inquire at La Pinta Hotel or at Rancho Santa Ines for a description of how to get to the prehistoric rock paintings (pictographs) located to the southwest in Arroyo La Bocana. These pictographs are known locally as "Cueva de las Pinturas Rupestres Gigantas" (*See* 3:162.5).

THE HISTORY OF MAN IN BAJA: The early history of man in Baja is not well known. The author of this guide has found, but not disturbed, many archaeological sites over the past 25 years. Scientific work is proceeding on the peninsula and on the Gulf islands. As one author wrote, even the islands remain "...largely an archaeological terra incognita." Extensive excavations and mapping of human habitation of the peninsula remains to be done to illuminate the history of human occupancy of Baja. Most of the investigations of human occupancy that have been done are either site reports or studies of artifact collections. The exact origins of the prehistoric Indians of the North American continent is unknown. One of the most popular theories of the origin of the American Indian states that millenia ago man came from Asia across the Bering Strait of Alaska either by way of the Diomedes Islands (whose formation served as a ladder) or the Aleutian Islands. It is thought that the various races of Indians arrived at different times and in several different migratory surges. All that is known for certain is that man has long inhabited the peninsula as evidenced by the numerous painted caves, petroglyphs, pottery shards, spear points, scrapers, manos and matates, debitage, carved bone, stone circles, rock cairns, lithic knives, shell middens, fragments of Yucca and Palm sandals, baskets, ceremonial effigies, and living structures.

Using stratigraphic and Carbon-14 data, some scientists calculate that the appearance of man in the Americas occurred between 10,000 and 35,000 B.C. Evidence indicates that man was in mainland Mexico about 11,000 years ago. It has been estimated that man first appeared on the Baja peninsula and some of the Gulf islands about 8,000 years ago. The only Indians currently living in Baja as a tribe are located at Catarina east of Ensenada.

NATIVE PALMS OF BAJA: Native palms were very important to Baja's Indians. They ate both the thin fleshy portion we normally eat and the large seed.

Palm fronds made numerous useful products such as sandal-like footwear, baskets, ceremonial effigies (dolls) of the dead, and house and roofing materials. There is evidence that the Indians also burnt the trees periodically to kill insects and mites and to improve the date yield or the next harvest. Palm berries are also eaten by birds and other animals. Orioles use the fibers from the leaf for nesting material.

Other vegetational dominants of the area are large Dodder-covered Elephant Trees, Barrel Cactus, Cardons, and Cirios.

171.5 The highway passes through a low saddle and descends into Arroyo La Bocana. Pools in or water down the arroyo can be seen most of the year. Downstream to the west on the "nose" of a large granitic mass, large, smooth, water-polished boulders are visible 20 feet above the wash bottom. At times there must have been at least 15-20 feet of water in this arroyo to have polished the tops of these enormous boulders.

It is possible to travel up nearby Arroyo La Bocana 23 kilometers to Mision Santa Maria. It was founded in 1767 by the Jesuit Priest, Father Victoriano Arnes, only a year before the expulsion of the Jesuits from Baja California. The mission is not badly deteriorated and is scheduled for restoration sometime in the future. If you do not have a 4-wheel drive vehicle it is worth spending some time in Santa Ines and having someone take you up the arroyo to Mission Santa Maria to explore this area.

This stretch of the highway passes directly over the site of the popular gas stop of old Rancho Catavina. The gas was siphoned into 5 gallon cans to measure it and then strained through a chamois or a felt hat into the gas tank.

174 This is Hotel La Pinta de Catavina. *Magna sin* is usually available.

174.5 Shortly after leaving El Pariador the highway crosses the Arroyo Santa Ines with its perennially-running, stream and another grove of native fan and blue palms. There is a picturesque view upstream of the rocks and another native palm oasis.

GAS STOP IN 1967

175.8 The paved turnoff to the left leads 2 Kms. to Rancho Santa Ines and a paved airstrip.

South of Arroyo Santa Ines the highway leaves behind the picturesque granitic rock of the Las Virgenes region and follows the old road up the arroyo. The hills are still covered by extensive dark lava flows.

177 A view of the unconformity with basalt over tonalite is visible to the left rear (*See* 3:164.).

179 This grade climbs through Miocene fluvial sedimentary rocks which cross the highway in a wide band. Fluvial sedimentary rocks are overlain by basalts similar to those on Llanos de San Agustin.

As the highway passes through the valley between volcanic hills, you will get an idea of just how rugged the old road, located in the wash about 100 yards west, used to be along this stretch of volcanic terrain.

Elephant Trees, Agave, Cardon, Cirio, Ocotillo, Mimosa (the bright green, thorny trees) also grow abundantly in this area.

182 At approximately 2 o'clock (south) on the lava-capped mesa, a small conical hill can be seen. It represents the eroded remnants of a small basalt cinder cone.

183 This section of the highway passes through granitic rock overlain by volcanic rocks. Several other conical-shaped hills, which also represents cinder cones, are in view on the horizon ahead.

184 White dikes cut across the mixed granitic and metamorphic rocks between here and Km 191.

186 A number of native fan and blue palms are growing to the right in the nearby Arroyo Jaraguay. The vegetation of this area is denser than that of other parts of the Vizcaino Desert because of better more moist soil. The vegetational dominants are typical of the Vizcaino Desert flora as represented by bright green Mimosa trees, Tamarisk, tall Cardon and Cirio, and a thick undergrowth of Garambullo, Pitaya Agria, Pitaya Dulce, Brittle-Bush, Creosote Bush, Ocotillo, Atriplex, Jumping Cholla, scattered Palo Estribo, and species of annual composites.

187 Rancho San Martin. The "golden spike" was planted here and commemorates the completion of the Transpeninsular Highway in 1973.

191.5 This turnoff leads southwest (to the right) to Rancho Jaraguay and the old road. Just before the ranch on the left are the ruins of a number of adobe buildings destroyed by weathering. In the old days this was a bath house where earlier Baja travelers could actually get a shower. This was unusual because there were not many places where travelers could even find drinking water.

Outcrops of gneisses and tonalite with andesite dikes are exposed along the highway. This region is one of impressive mixtures of metamorphic and granitic rocks, dikes, bouldery outcrops, and beautiful basalt-capped mesas such as Mesa Jaraguay, Mesa El Gato, and Mesa Prieta.

193 The highway climbs a steep grade and crosses over the path of the old road.

The compass plant of Baja (Barrel Cactus) is abundant to the left of the highway.

194.5 As the highway crests the top of Jaraguay grade (823 meters), a good view opens to the north (rear). This viewpoint provides an overview of the volcanic tableland the highway has been passing through, the unconformity between the granitics and the basalts, and the relatively flat erosion surface that underlies the volcanic strata.

A few Cirios are growing on the drier south side of the grade, and many more larger Cirios can be seen on the moister north side.

NORTH AND SOUTH FACING SLOPES: Local variations of humidity and temperature exercise considerable control upon natural slope cover. In nontropical regions, slope face is of great importance. Because Baja is north of the Equator, its southward-facing slopes receive more direct sunlight, and they are hotter and drier. As a result of the decreased humidity and increased temperatures, slope cover is sparse, low, and composed of grayish-colored species which are adapted to living in xeric (dry) environments.

North facing slopes receive less direct sunlight and are more humid and cooler than south facing slopes. They support denser, lusher, taller, moisture-needing vegetation including trees characteristic of mesic (moist) environments. Wind is slowed, and soil temperatures and evaporation are decreased by the dense vegetation. The soils are less weathered and are humus-rich.

195.5 At this kilometer a good view of Laguna Seca, which is normally a dry lake, is visible to the left. The old road used to bear to the left and crossed part of the lake bottom. At the present time the highway circles around to the right side of the lake on the high ground near the base of the hills.

On John's first trip down the peninsula, he and his wife Carol were slowly approaching Laguna Seca on a very hot afternoon. That day the heat waves made the dry lake look like it was filled with water. As he looked at the Lake, he was amazed to see an 18 foot sail boat. He exclaimed to Carol, "There's a sail boat on that lake!" Carol felt his forehead and asked cautiously, "Do you feel all right?" However, he was right; there was a boat on the dry lake. Someone had tried to trailer a sailboat up the peninsula and had gotten this far before all the springs on the trailer broke. Arturo, at Rancho Chapala, talked about the crazy Gringo for years.

SAIL BOAT ON DRY LAKE

198.5 A view opens to the south. Several volcanic peaks of the San Jose volcanic fields can be seen in the distance to the right. The rolling landscape in this region is covered with volcanic debris.

202.5 The highway crests over a rise and reveals the first good view of Cerro Pedregoso.

PEDREGROSO

206.2 The highway passes close to Cerro Pedregoso (which means stony or rocky). This granitic hill has been covered by volcanic and other debris so that only the top stands out like an iceberg above the volcanic plain (Pliocene-Holocene basalts and basaltic andesites). The granitic rocks have been spheroidally weathered which left large boulders of granite which look like they were piled on the hill.

A dark andesite dike cuts through the middle of the hill. Cerro Pedregoso, a very prominent feature on the otherwise fairly barren, gently rolling landscape, acted as a beacon for Baja travelers could see for many kilometers as they slowly approached the hill over the old rugged dirt road.

215.9 This roadcut exposes andesitic dikes in tonalite.

In the late spring the tall "weeds" with yellow-centered white flowers which grow along the roadside are called California Prickly Poppies.

217 The red and black pyroclastic cinders and basalt exposed in this roadcut indicates that an ancient eruptive center is located nearby.

218 A prominent white quartz dike is exposed approximately 100 meters to the left of the highway. The old road passed by that dike.

221 The hill ahead is cut diagonally by a series of prominent, gray-white, granitic dikes.

222.7 One of the dikes noted at Km 221 forms the ridge approximately 100 meters to the left of the highway. This dike is approximately 7 meters thick and is very light in color.

GRANITIC DIKES

223 The highway begins a gentle descent as the dry playa lake of Laguna Chapala comes into view. Outcrops of volcanic rocks overlying granitic rocks are exposed along the highway.

224 The mountains to the southeast are part of the Sierra la Asamblea. The valley in front of them is Valle Calamajue.

225.5 Several more prominent white granitic dikes cut the hills to the right front of the highway.

228 At this kilometer there is an excellent view of Laguna Chapala. In the distance to the left is the main part of the dry lake bed playa.

DUSTY CHOICES AT LAGUNA CHAPALA

The old road went to the left across the lake to a small group of buildings and trees. The dying trees, which are barely visible on the far side of the lake, are all that is left of the old Rancho Chapala. Now, the highway skirts the west side of the lake and passes between the upper and lower lake at the New Rancho Chapala.

Pulvo = Dust: This was one of the dustiest parts of the old road on the entire peninsula. The heavy traffic across the lake on the old road pulverized the dry lake bed sediments into a choking dust. Two former students of mine once crossed this area in an open jeep. One of them stood on the seat and held onto the windshield while he "almost" directed the driver through the ruts! The driver high centered the jeep, and they had to jack the jeep up in 10 inches of dust. There was no "best route". At the flatter south end of the dry lake, it was possible for a vehicle to achieve speeds of 60 mph! For a mile or two.

At Laguna Chapala a vegetation change can be seen. The Cirio almost disappears; Cardons become very small against the foothills of Sierra de Calamajue. The vegetation adjacent to the highway is dominated by the halophytes, low gray mounded Bursage, and Creosote Bush. Occasional Datilillo, Cholla Cactus, Pitaya Agria, Desert Mallow, and Teddy Bear Cholla are scattered among the halophytes.

229.5 The roughly graded unpaved dusty road to the left leads to the northeast to Bahia de San Luis Gonzaga, Puertocitos, and San Felipe. Preparations are being made to pave the portion of the road between Laguna Chapala and Puertocitos. It is not recommended that passenger cars attempt this road until it is paved. (*See* Log 11).

230.8 The new Rancho Laguna Chapala.

TILTING OF THE PENINSULA? The old shorelines of Laguna Chapala are tilted relative to the Recent shorelines of the lake. This indicates Pleistocene to Recent tilting of this area.

232 The semi-stabilized dunes on the south end of Laguna Chapala are blown from the lake during normal dry spells. A climb to the top of the dunes will provide an excellent view of Laguna Chapala.

DRY LAKE BED OF LAGUNA CHAPALA

234 The hills ahead of this kilometer mark are composed of tonalite which is cut by dark andesitic and basaltic dikes which fed the extensive basaltic volcanic field which form the hills to the right.

235.7 Excellent examples of black basaltic dikes cutting granitics, are exposed in this roadcut.

The taller vegetation along the highway is composed of Mimosa and Indian Tree Tobacco.

INDIAN TREE TOBACCO is the common name of the tall (1-3 meter) herb which grows in the disturbed soils along the roadsides throughout the peninsula. This plant is easily recognized by its yellow tubular flowers and disagreeably strongly-scented poisonous narcotic leaves. Indians used the plant to cure ailments of the chest and lungs and as a headache and toothache remedy. Its juice was used to promote wound healing, and the dried leaves were smoked for their narcotic effect. This plant is reported to be poisonous to stock. Some Indians died from its use.

238 The old road joins the highway from the left just before passing through Cuesta El Portezuelo.

Good stands of vegetation typical of the Vizcaino Desert return. The flora is dominated by Cirio, Cardon, Ocotillo, Elephant Tree, Barrel Cactus, Old Man Cactus, *Mammalaria*, and Cholla.

239.5 At this point the highway passes Cuesta El Portezuelo. A small shrine is located to the left on the old road below. This is the main peninsular divide where the highway passes east of a ridge between the Pacific Coast drainage and the Gulf drainage.

The highly mineralized granitic and metamorphic rocks of Sierra la Asamblea form the high mountains to the left front. The high light-gray part of the Sierra is a granitic pluton which is surrounded by metamorphic rocks. Several light-colored dikes cut the darker metamorphic rocks.

SIERRA LA ASAMBLEA

242 The vegetation of this area is similar to the granitic boulder gardens which surround Catavina (*See* 3:153.5) and is dominated by Cardon, Cirio and Elephant Trees, Old Man Cactus, Teddy Bear Cholla, Yucca, Agave, Jumping Cholla, Pitaya Agria, Palo Adan, and Ocotillo.

248.5 There is a view of Cerrito Blanco directly ahead and to the left of the highway. It is a low light-gray granitic hill, which is highly jointed. The jointing makes the hill look like it is composed of bedded sedimentary rocks.

250 Granitic and metamorphic rocks form the low hills to the east of the highway.

251 The light-colored flat material along both sides of the highway for the next several kilometers is Pleistocene lake-bed sediments.

253 The low white hill to the left is Cerrito Blanco. This jointed granodiorite hill sits on an alluvial plain which is unconformably overlain by white cross-bedded Paleocene sandstones.

To the north is a view of a volcanic tableland and the cinder cone peak of Cerro el Volcancito.

256 Datilillo, Cardon, and Cirio have are the dominant tall plants of the Vizcaino Desert flora.

257 The highway crests a very low divide and moves back into the Pacific slope drainage.

258.5 The old road turns off to Bahia de San Luis Gonzaga, Puertocitos, and San Felipe. This road can be used as a route to reach the newer previously mentioned graded dirt road from Laguna Chapala at Km 229.5. This road passes by Mision Calamajue, Cerro el Volcacito, through Valle Calamajue between Sierra la Josefina and Sierra Calamajue, and connects with the graded road to Bahia de San Luis Gonzaga at 11:21.3.

Several more raptor nests can be seen in Cirios along of the highway. Woodpeckers commonly excavate holes (nests) in the Cardons in this area. Woodrat nests can also be seen in the centers of yucca clumps.

CIRIO - OCOTILLO - CREOSOTE

262.1 This small grade represents a prominent fault scarp that cuts diagonally across the highway. The fault uplifts Miocene fluviatile sedimentary rocks on the south against the Quaternary alluvium and lakebed sediments on the north.

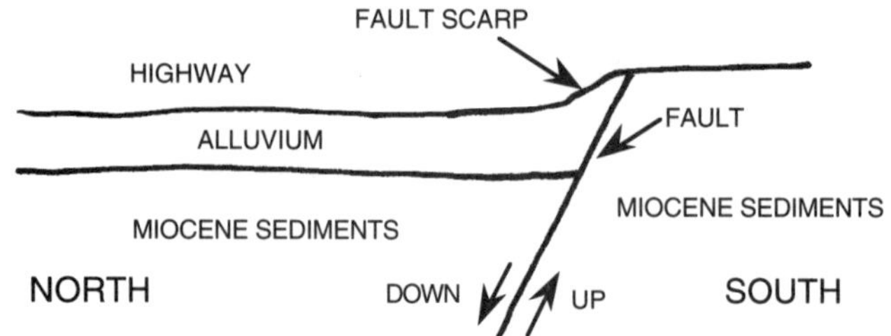

WOODPECKER HOLE IN CORDON

VIZCAINO PHYTOGEOGRAPHIC REGION FLORA

This is the north edge of the stable San Borja block (Gastil *et al.*, 1975). It consists of a pre-Miocene, west-sloping bedrock surface, which is discontinuously overlain by Cenozoic sedimentary and volcanic strata. South of here the uninterrupted mesas extend from the main Gulf escarpment to the Pacific coastal plain.

265 The highway begins a gradual descent into the drainage system of Arroyo el Crucero and follows it for several kilometers. This drainage is developed on and eroded into Paleocene marine and nonmarine sedimentary rocks.

Just north of Punta Prieta the dominant species alternate between Cirio, Datilillo and Cardon. The low sparse gray "understory" is Bursage (*Franseria dumosa* and Agave, Garambullo, Jumping Cholla, and Acacia. This Datilillo-Cardon/Datilillo-Cirio community continues for many kilometers.

HOW DO THE LOW SPARSE GRAY FLORA OF BAJA'S EXTREMELY ARID DESERTS SURVIVE? Most of Baja's desert landscape is dominated by low-growing vegetation (shrubs, cacti, and grasses) although some plants do grow to the size of small trees. All of these plants are remarkable in the many and varied ways in which they have adapted to a desert existence. Some are leafless (obligate drought-deciduous species) most of the year. In the absence of leaves, the green trunks (cladodes) act as photosynthetic leaves and produce carbohydrates to meet the plants required energy resource.

Others have silvery or velvety foliage that helps reflect heat and retain moisture under the hot desert sun. Many are armed with spikes or thorns, perhaps to discourage animals from grazing on their slow-growing branches. The spiked leaves of many plants provide points from which dew can collect, fall to the ground, and water the plant. Other adaptive features are the brief life span of some plants which germinate, grow flowers, and produce seeds in days to a few weeks; the storage of water in tissues and the accordion structure of cacti which swell to hold more water. Some have very large trunks like the Elephant Trees or develop thick and waxy outer skins. Wide spacing, shallow roots, and vegetative re-production are also water saving adaptations.

280.5 Junction of Highway 1 and Highway to Bahia De Los Angeles (*See* 16:0), the Kms. change to 0.

Log 4 - Bahia de Los Angeles Turnoff to Guerrero Negro [129 kms= 80 miles]

***L. A. Bay Turnoff to Rosarito -** The Highway continues south in the Miocene fluvial sedimentary rocks with basalt mesas to the left and high rugged metamorphic and granitic hills to the right. It then drops into the main alluvial channel of Arroyo Leon, past Punta Prieta, with Paleocene fluvial sedimentary rocks on both sides of the arroyo and metamorphic and granitic hills in the distance.*

At La Bachada, the Highway climbs up a grade through rolling hills onto the rolling Paleocene mesas with isolated steep hills of metavolcanic rock and high volcanic mesas in the distance. After passing a rugged metavolcanic hill to the left the two conical shaped Occidental Buttes composed of Paleocene fluvial sedimentary rocks are seen to the right. The Highway then enters a hilly area of granitic rocks and drops into El Rosarito.

***Rosarito to Guerrero Negro -** The Highway follows the south bank of the arroyo with metamorphic rocks on the left, then passes through a canyon in the steep gabbro hills before turning south to again climb onto and through rolling hills in the undulating dissected mesas of marine Paleocene sedimentary rocks. The Highway drops into a steep narrow gorge in gabbro then alternately climbs onto the Paleocene marine mesas and drops into the valleys with the basalt capped Paleocene mesas to the east.*

The Highway then drops onto the alluvial fans of the Llano del Berrendo. During the long crossing of this plain the low hills of metasedimentary rocks to the left become more distant. A low basalt cone is closely passed to the right and the distant basalt cone of Punta Santa Domingo looms ever closer. The coastal dune field and occasionally the Pacific Ocean are in almost constant view to the right. After passing Jesus Maria the Highway begins to cross the limestone surface of the Pleistocene lagoon. The dune field encroaches on the Highway as the Eagle Monument is approached at the state line.

0 The highway to the left leads to Bahia de Los Angeles (*See* log 16) while the highway ahead continues 129 kilometers to the southwest to Guerrero Negro.

4 The flat mesa to the left front is Miocene fluvial sedimentary rocks capped by basalt.

6.5 The Cardons which are seen along this stretch of the highway are reputed to be among the tallest in Baja, even taller then some specimens in the large cardonal located at the south end of Bahia de Concepcion on Baja's gulf coast.

GIANT CORDON

9 The dominant trees along the highway are two unrelated genera of Elephant Trees, *Pachycormus* and *Bursera.*

THE NAME ELEPHANT TREES is unfortunately applied to several unrelated desert trees in Baja. The two kinds of Elephant Trees in this region are not closely related at all. *Pachycormus discolor* is a member of the Cashew family, while *Bursera microphylla* belongs to the Torchwood family. These two Elephant Trees are easily differentiated because of the distinctive incense-like aroma which is given off by the crushed leaves of the *Bursera.*

ELEPHANT TREE

Bursera microphylla ranges from the Anza-Borrego Desert in the U.S.A. through the entire desert regions of the peninsula to the Cape region. A second species of *Bursera, B. hindsiana*, has almost the same distribution. Oil from the fruit of *Bursera* has been used as a dye and for tanning hides. A fourth species of Elephant Tree, *Bursera odorata* may also be seen.

13.2 This turnoff leads to Punta Prieta (Named for the dark basalts seen in this region). The main highway follows Arroyo Leon which is developed along a fault that is responsible for the arroyo's north-south trend. The Paleocene Sepultura Formation is exposed on both sides of the arroyo for the next 10 kilometers (*See* 4:29 & 4:44).

14 The airstrip at Punta Prieta is to the left. The hills to the right are composed of metavolcanic rocks of the Cretaceous Alisitos Formation. The Paleocene sedimentary rocks lapped against the irregular topography of the Cretaceous hills. The high hill to the east is composed of tonalite.

16.5 The denser vegetation in this region continues to be representative of the flora of the Vizcaino Desert. The plants along the highway are predominantly Cardon with occasional Palo Verde, tall lichen-draped Cirios, Agave, infrequent shorter Barrel Cactus, Datilillo, Pitaya Agria, Pitaya Dulce, two species of Elephant Trees in the washes, Jumping Cholla, Teddy Bear Cholla, Creosote Bush, Broom Baccharis, Cheesebush and the last scattered occurrences of Ocotillo. From this point to the south an Ocotillo look-alike, Palo Adan, will begin to replace Ocotillo.

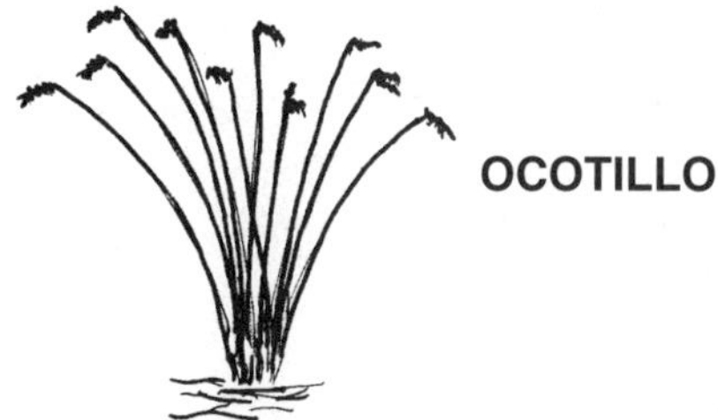

OCOTILLO

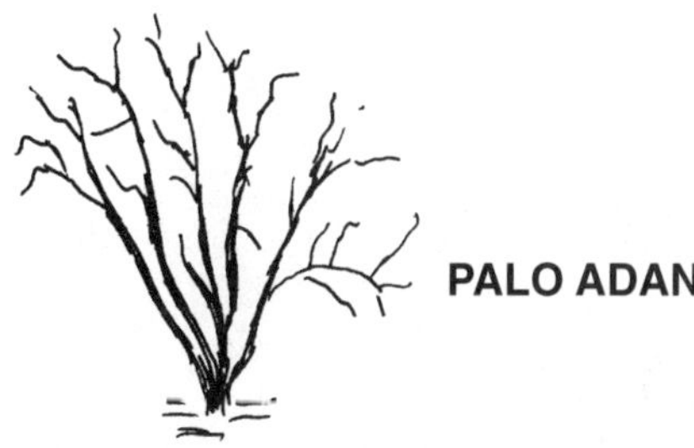

PALO ADAN

OCOTILLO OR PALO ADAN? These two plants along with the Cirio belong to the Ocotillo family, a group of shrubs with long erect thorny whiplike branches. It may seem like Ocotillo (*Fouquieria splendens*) and Palo Adan (*Fouquieria diguetii*) are the same plant. The characteristics that help separate these two close relatives are general trunk, branch, and flower morphology (external appearance); geographical distribution; and habitat preference. The chart below will help to distinguish between these two look-alike relatives.

	Palo Adan	Ocotillo
Trunk	short and thick	none
Branch diameter	Thick; branches off a short trunk	Thin, slender, whiplike, spreads upward, fanlike from ground
Flowers	Panicles smaller	Panicles larger
Geographical distribution	Abundant from the central part of the peninsula south to the Cape	Abundant from California to just north of Guerrero Negro rare south to the Cape
Habitat preference	Clay and granitic soils of alluvial plains	Desert slopes and plains

OCOTILLO

PALO ADAN

24.5 The highway turns and crosses Arroyo Leon where a wash woodland is dominated by Mesquite and scattered Datilillo.

25.5 This is Rancho La Bachada whose well was once one of the better sources of water in this part of the peninsula. In 1965 the San Diego State University geologic mapping crews were based at Punta Prieta. Once a week someone was sent to La Bachada to get 100 gallons of drinking water from the well. One week a crew member was sent to fill eighteen 5-gallon jugs. As he was filling the eighteenth jug, he lifted a bucket of well water which contained a dead decaying hairless rat. Not wanting to go elsewhere to refill the jugs, he pitched the rat, drew another bucket of water to fill the last jug, and returned to camp without saying a word (he used the other two 5-gallon jugs that week). No one found out what had happened until years later.

26 The highway begins a steep ascent through Paleocene conglomerates at the northernmost edge of the Paleocene embayment.

29 At the top of the grade, the vista opens to the south and presents a view of the Pacific coast and the flat Paleocene Sepultura Formation which forms all of the flat-topped mesas in the immediate foreground and into the distance. The rugged high peaks to the far right and the high sharp peak directly ahead are composed of metasedimentary rocks of the Alisitos Formation.

This area was once a Paleocene marine embayment. Shallow water marine to nonmarine sedimentary rocks from this embayment have yielded fossils of hackberry seeds, marine invertebrates, a tooth of the proto-horse *Hyracatherium*, and nonmarine vertebrate fossils (by Occidental College).

THE PALEOCENE ENVIRONMENT: Fife (1968) reconstructed the probable Paleocene environment and stated, "The coast north of Punta Santa Rosalia... was rocky as it is today. The location of fossiliferous deposits indicates that heavy-shelled gryphoid oysters were deposited with conglomerates near shore; while turritellas and echinoids were preserved with finer clastic sedimentary rock in the shallow embayments..." "South of Punta Santa Rosalia at least three major embayments existed..." "The southern embayment is well exposed and contains shallow to brackish water faunas. These were areas of mud to sand bottoms. The position of fossil mollusca, corals, ostracodes, and foraminifers suggests that alternate shallow marine and estuarine or brackish water conditions existed. Seeds of the Family Chenopodieaceae and the remains of the proto-horse *Hyracatherium* are found in beds interfingering with typical estuarine strata. Ostracodes and charophytes indicate a lacustrine environment in the continental beds east of Rancho La Bachada."

THE ORIGIN OF THE HORSE AND ITS PLACE IN BAJA TODAY: It is hard to believe, but horses evolved in North America. Modern domestic horses (*Equus caballus*) belong to a small order of mammals known as the Perissodactyla or odd-toed ungulates which first appeared in the late Paleocene (58 million years ago) in North America. The earliest of the horselike ancestors *Hyracatherium* appeared in the Eocene about 54 million years ago. *Hyracatherium* was a small dog-sized mammal that browsed on low shrubs of the forest floors. It had already lost two hind toes on its hind feet and one on its forefeet, but the feet were still covered with soft pads. Its the tooth of this horse ancestor that was collected from the Paleocene Sepultura Formation.

When grasses appeared in the early Miocene, hypsodont grazing equids appeared, and by the late Miocene equids had reached their peak of diversity. The need to run from predators and to travel long distances in search of food and water led to many changes in equid body shape including increased body size. By the early Pleistocene, two million years ago, the one-toed equids (*Pliohippus*) had spawned the genus *Equus* which spread rapidly over the world. The center of equid evolution was North America. True equids did not migrate to the Old World from North America until the early Pleistocene about two million years ago. The Old World equids evolved from the three-toed North American immigrant equid *Hipparion*. By the end of the Pleistocene modern equids had become widely distributed on all continents except Australia. For anunknown reason all North American ungulates, except the pronghorn antelope, became extinct about 10,000 years ago. In the Old World the modern horse, *Equus caballus*, was first domesticated in Asia about 5,000 years ago.

The domestic horse was reintroduced into North America (mainland Mexico) by the Spanish in the early 16th century.

32.5 This hogback ridge is composed of complexly folded and overturned metavolcanic rock of the Alisitos Formation. This unit is exposed to the left for the next 4 Kilometers.

39 The small palm oasis to the right is known as Agua de Refugio. The waters of the small stream originate from springs located to the left of the highway.

39.5 The highway crosses a small heavily vegetated wash with a perennial stream which originates in the metavolcanic hills to the left.

44 Occidental Buttes, to the right, are two small buttes in the Paleocene Sepultura Formation. This is the site of vertebrate fossil discoveries by an expedition from Occidental College in the mid-1960's. Mother nature is taking a sun bath.

OCCIDENTAL BUTTE

PALEOCENE SEPULTURA FORMATION: "The most extensive outcrop of fossiliferous Paleocene marine rocks occurs in the southern part of the area, in the vicinity of and to the north of Rancho San Xavier. In this region at least one hundred feet of dusky yellow to reddish brown sandstone is interbedded with conglomeratic sandstones, concretionary lenses, siltstone, and mudstone. Several horizons, forming resistant strata, contain *Turritella pachecoensis* almost exclusively. Locality B5D-30 yielded *Cerithidea sp.*, *Ostrea sp.*, *Venericardia sp.*, *Glycymeris sp.*, and *Turritella pachecoensis*, plus several specimens of gryphoid oyster and unidentified horn corals, ostrocodes, and foraminiferas. This assemblage represents a sublittoral facies. Farther north, about six miles [ten kilometers] up the arroyo from El Muertito, the assemblage suggests a lagoonal or littoral environment. At this location the remains of a small proto-horse, *Hyracatherium sp. nov.*, were interbedded with seeds from the family *Chenopodiaceae, Cerithidea sp.*, *Calyptraea sp. nov.*, and *Ostrea sp....*" (Fife, 1968)" About six miles due north of the above locality, at Occidental buttes, Morris (1966) reported the discovery of ungulates of the Orders Tillondontia, Perissodactyla and Pantodonta. Pantodonts of the family Barylambdidae were found stratigraphically above specimens assigned to Tillodontia and Perissodactyla..." (Fife, 1968).

45 The vegetation of this region continues to be dominated by species characteristic of the flora of the Vizcaino Desert, and Cirio is the predominant tall plant. Plants associated with these Cirio are Barrel Cactus, Elephant Trees, Purple Bush, Agave, Pitaya Agria, and an occasional Cardon, Palo Adan, and Jumping Cholla. The common epiphytic lichen, *Ramalina reticulata*, can be seen growing on the Cirio and Elephant Trees.

46 The highway climbs through conglomeratic beds and begins a descent through redbeds of the Paleocene Sepultura Formation.

47 The high mesa to the left of the highway is the lagoonal section of the Paleocene Sepultura Formation overlain by basalts.

49 In addition to the lichen the Ball Moss (*Tillandsia recurva*) is growing epiphytically on the Cirio and Elephant Trees.

EPIPHYTES AND PARASITES: From south of El Rosario along the Pacific coast to the Cape Region, trees, shrubs, and cacti such as Palo Adan, Cirio, Lomboy, Broom Baccharis, Cardon, and Pitaya Dulce are often abundantly covered with the moisture-loving epiphytic Ball Moss, a member of the pineapple family. Epiphytic Ball Moss is not harmful to the host plant, since it lives in a commensalistic relationship which depends only on the host plant for support. However, a close look at the same trees, shrubs, and cacti listed above will reveal harmful parasites which live in a destructive symbiosis with them. The two common parasitic organisms of this area are the evergreen mistletoe (*Phoradendron californicum*) and "witches hair" (*Cuscata veatchii*).

"Witches hair", a plant lacking chlorophyll, attaches its yellow or orange hairlike stems to its host by a modified stem called a haustoria. Using the haustorial stem, the "Witches Hair" steals sugar produced photosynthetically by its host.

The second parasite, the evergreen mistletoe, hangs in trailing shaggy strands from many plants in Baja. Mistletoe also utilizes haustorial stems to drain sugar from its host. The moisture that sustains both the epiphytes and parasites of this desert comes from ocean fogs (Neblina) that roll in at night and lies during the early morning hours in misty layers among the hills until the sun "bakes" it off. Along the Pacific from the Cape north to San Francisquito, similar areas are known as coastal fog deserts. Mistletoe is less common on the drier Gulf side of the peninsula where fogs do not occur.

52.5 The main part of the small village of El Rosarito is located to the left.

A rough road from the center of town leads east 32 kms through very dense vegetation to Mision San Borja. The mission was completed in 1762 with money supplied by Maria, the Spanish Grand Duchess of Borja. It was abandoned in 1818, but it has been restored. It is a beautiful place to visit with an off-road vehicle.

MISION SAN BORJA

The wash which roughly parallels the highway is vegetated by green grasses, rushes (*Juncus sp.*), and scattered native palms.

55 Datilillo, Cardon, Cirio and Elephant Trees are abundant; however, most of these Cirio specimens are rather short when compared with some of the "giants" of the moister habitats to the north.

After crossing the arroyo the low hills to the right of the highway are gabbro. The hills ahead and a bit to the left are prebatholithic slates.

TIGHTLY FOLDED SLATES

56.8 The highway turns south out of the arroyo. Granodiorite is exposed to the right in the lower end of the Arroyo El Rosarito.

59 Along this stretch beds of the Sepultura Formation are exposed on both sides of the highway. The low hills to the east are schist and gabbro.

60 This is the first view of the ocean since El Consuelo (9 kilometers north of El Rosario).

62.7 Turnoff to Playa Alta Mira on the Pacific coast.

64 Good exposures of the marine red beds of the Sepultura Formation are in the arroyos.

66 The highway descends into a gabbro pluton exposed in the narrow neck of a canyon and then climbs through the reddish beds of the Paleocene Sepultura Formation.

The southernmost Cirios are growing here with Elephant Trees, Candelabra Cactus, Pitaya Agria, Palo Adan, and Creosote Bush.

68.5 The highway descends into the broad arroyo. The prominent red flat-topped hill directly to the left is a gabbro-serpentine pod. After it climbs out of the arroyo, the highway passes through more outcrops of the Sepultura Formation.

The graded road to the right leads to El Tomatal Beach and Miller's Landing. Onyx from the abandoned mines at Marmalito was once shipped from the white-water beaches of El Tomatal.

70 For the next 60 kilometers the highway travels south along the flat plain north of Guerrero Negro providing panoramic views of flat-topped mesas to the east and the blue waters of the Pacific to the west. The prominent dark-topped mesas to the left are basalt overlying lacustrine and lagoonal rocks to shallow water marine beds of the Paleocene Sepultura Formation. To the right in the distance, a volcanic hill forms the north end of Laguna Manuela.

The highway continues to traverse a relatively flat gently undulating plain for several kilometers, occasionally dropping into and through numerous small arroyos. Then it crosses the 150 Kilometer expanse of Llanos de Berenda. It is referred to as Antelope Plain because of the once abundant herds of antelope. At the present time there are only few antelope left.

The grasslands of the world are the homes of most of the great herds of herbivores. In Baja the Pronghorn Antelope (*Antilocapra americana*) is the only large feral grazing ruminant which inhabits the vast Llanos de Berenda. The herds of pronghorn antelope have decreased due to hunting pressures (now illegal in Baja) and the destruction of the grasses the antelope feed on as this region was changed by agriculture and cattle-grazing practices. Today they have retreated to rougher higher country unsuitable for agriculture or cattle-grazing. Although they look like deer, the pronghorn antelope are not members of the deer group, nor do they even have much connection with the Old World antelopes. The horns of the pronghorn antelope have a permanent bony core which is not discarded, although the outer horny modified hairlike covering (keratinized skin) is shed once a year. Old World antelope never lose any part of their entirely bony horns, and they lack the outer horn sheaths.

80 The low dark hill to the right is a very small basalt cone (*See* 4:98). The point of Guerrero Negro comes into view in the far distance.

The dune fields to the right parallel the Pacific coast and extend from the edge of Laguna Manuela to the north end of Scammon's Lagoon.

This **sand dune building** has resulted from the southward transportation of sand along the coast by long shore currents. The currents have caused a sand bar to build south of Punta Manuela. The winds pick up the sand and carry it into the lagoon and landward away from the coast. This builds a set of migrating coastal dunes both on the bar and on the landward side of the lagoon. As the dunes progress inland, they move farther from the supply of coastal sand and soon become stabilized by plants which utilize the moisture trapped in the sand. The vegetation between these migrating dunes is quite sparse and consists of a few salt tolerant species (halophytes) such as Pickleweed and Sand Verbena. The tops of the well drained less saline soils of the few stabilized dunes support greener specimens of Mesquite, Atriplex, and Ice Plant.

A good place to see this relationship is at the old barge loading area west of town. The barge loading area was moved into Laguna Guerrero Negro, because navigation was hindered by the drifting sand that built sand bars in the lagoon.

In this region most of Llanos de Berenda is underlain by Pleistocene limestones deposited when this part of the peninsula was covered by a warm shallow sea. Limestone is alkaline and makes it harder for plants to establish themselves. As a result, the Llanos de Berenda is quite sparsely vegetated by the Palmer's Frankenia/ Datilillo plant community with scattered Cholla.

A **plant community** is a regional assemblage of interacting plant species characterized by the presence of one or more dominant species. Some plant communities are named for the tree or shrub species which are dominant in them. The term "dominant" refers to one or more plant species which may be the largest or most abundant species in a community. Because of the foliage cover or the extent of their root systems, dominants have a strong influence on the local ecology of the plant community.

82.5 The highway passes between Rancho San Angel and a low hill of basalt to the right. The arroyos adjacent to the highway are seasonally brightened by flowering Agave (Maguey), Broom Baccharis, Tamarisk, and Barrel Cactus.

LIVING FENCES: The fence posts along this section of the highway and in many other parts of Baja are made of the cut branches and trunks of several cactus species, primarily those of the genus *Pachycereus*—the Cardon, the Galloping Cactus or Pitaya Agria (*Machaerocereus*) and a member of the Lily family, such as; the Datilillo, (*Yucca valida*).

DATILLO

After the plants are cut and placed in the ground for fence posts, they re-root and continue to grow. It is possible to asexually reproduce many plant species by placing cuttings of their stems or branches in moist soil. The cuttings develop adventitious roots and become whole new plants.

Numerous desert plants in Baja reproduce in this fashion. For example several species of Baja's Cholla cactus have pads which readily fall from the main plant, develop adventitious roots, and become new plants. These plants are said to have propagated by vegetative asexual reproduction. Other species which reproduce in this manner include Cardon, Yucca, and Datilillo. Vegetative reproduction allows desert plants to avoid the costly water and energy-consuming processes of sexual reproduction and permits the spread of plant species even when its too dry for seed production and the establishment of new plants by seed germination.

83 A view to the left rear shows a reddish cinder cone on top of the prominent dark mesa.

98 The dark peak at 3 o'clock is a basalt cone forming Punta Santo Domingo at the north end of Laguna Manuela. It has a shield-like shape with a prominent, steep, dark, eruptive cinder cone in the center. The gentle slopes are lava flows, while the steep central slope is a cinder cone.

99 The road metal quarries on the left side of the highway expose the marine Pleistocene limestone which covers large parts of the Vizcaino Peninsula south of and the Magdalena Plain.

115 Almost nothing over 0.3 meters tall grows from here to Guerrero Negro. The low gray (10-30 cm) tufted mound-sloped plant which covers the Llanos de Berenda is a bursage, Palmer's Frankenia (*Frankenia palmeri).*

PALMER'S FRANKENIA is a stiff woody shrub which grows on sandy areas, salt marshes, and alkali flats throughout the peninsula. The white flowers, which bloom in November brighten the entire region around Guerrero Negro.

A walk through the ground cover of Palmer's Frankenia reveals epiphytic foliose lichens and extensive algae mats that nearly cover all of the bare ground between the Bursage. Algae mats such as these are nitrogen fixers which add valuable nitrogen compounds to the desert soils.

PALMER'S FRANKENIA

A taller gray-green leafed introduced Australian weed commonly known as Australian Saltbush (*Atriplex sp.*) grows near the highway shoulders or on the sand dunes. This weed, now a aturalized "native", was originally cultivated for cattle forage but escaped and now grows extensively in saline soils and along the shoulders of the Highway throughout the peninsula.

118 As the highway continues southward , a 40 meter high metal eagle monument, "Monumento Aguila", dominates the highway. This monument was built on the 28th parallel to commemorate the completion of Highway 1 which joins the states of Baja California and Baja California Sur.

120 The highway approaches and then crosses a semi-stabilized dune field.

123 The coastal lagoon of Laguna Manuela will be in view to the right of the highway for the next half kilometer.

128 = 220.5 This is the State Line eagle monument at the 28th parallel and Hotel La Pinta de Guerrero Negro. The kilometer markings will now descend to zero points.

The 28th parallel marks the boundary of the Pacific Time Zone (to the north) and Mountain Time Zone (to the south).

OSPREY NEST

OSPREY IN FLIGHT

For years there has been a huge nest of sticks belonging to a pair of Ospreys (one of the predatory birds discussed at 3:147) that have nested on the sign on the east side of the monument.

Ospreys or Gavilán Pescador or "Fish Hawks" breed along both coasts of Baja and the western coast of mainland Mexico (Sinaloa and Sonora). They are easily recognized by their clear white underparts and black "wrist" marks in flight. From their large nests the Ospreys fly to the Pacific and circle overhead or hover with tail spread and feet dangling while watching for the gleam of a fish near the surface. From a height which may reach a hundred feet, the "Fish Hawk" dives into the water and emerges with it talons impaled in the back of a fish. They build bulky nests in trees, on poles, and docks.

EAGLE MONUMENT

216.5 The highway forks here. The left fork leads to San Ignacio; the right fork leads 2 kms through supertidal flats, which are intermittently flooded by very high tides, into the town of Guerrero Negro. To see the birding and whales at the barge loading area, go into town, cross a channel and turn right at the main cross street after the bend.

THE WORLD'S LARGEST SALT PRODUCING REGION: Over 5,000 people live in the town of Guerrero Negro. The town's name "Black Warrior", refers to a whaling bark (ship) wrecked in the lagoon in 1858. According to Exportadora de Sal, the salt company which owned the town, the economy of Guerrero Negro is supported by the "world's largest" salt-producing operation.

Salt is a general term for naturally occurring sodium chloride (NaCl), a substance essential to life. Fortunately, it is one of the most abundant substances in the world. Salt is produced commercially from brine, salt beds, salt pans, and salt domes world wide. To the south of the highway on the vast tidal flats of Laguna Ojo de Liebre (Scammon's Lagoon), are man-made salt pans which cover over 300 square miles. Exportadora de Sal S.A. have diked some of the more shallow parts of the tidal flats of Scammon's Lagoon forming shallow evaporating ponds which are approximately 100 meters square and one meter deep when flooded with sea water. As the intense sun evaporates the sea water, the sulfates and carbonates precipitate out leaving a salty brine solution that is moved to adjacent salt pans to precipitate the halides (table salt, NaCl, is the most common). When the brine has completely evaporated, the salt precipitate is collected, loaded into triple trailer bottom-dumping trucks (150 feet long), and transported to the loading docks southwest of town. At the loading docks salt is loaded on barges and lightered to Isla Cedros where it is shipped to the United States, Japan, Canada, and Mexico. It has been estimated that over 5,000,000 tons of salt are shipped throughout the world. Guerrero Negro is not the only salt-producing site in Baja. Another 50,000 tons are produced annually at the solar salt works located on Isla del Carmen.

GUERRERO NEGRO SALT MARSH

GUERRERO NEGRO SALT WORKS FROM AIR

HOW SALT CAME TO MEAN SALARY: The word "salt" comes from the Latin word "sal". During the time of the Roman Empire, soldiers were paid partly with the coin of the empire and partly with salt. If a soldier failed to fully perform his duties, his ration of salt would be cut or withheld because he was "not worth his salt". The English term "salary" was derived from the custom of partially paying workers with salt. Today salary refers to a fixed payment received at regular intervals for services rendered.

WHALE VIEWPOINT IN LAGOON

BORREGO IN CUEVA PINTADA
SIERRA SAN FELIPE

Log 5 - Guerrero Negro to Santa Rosalia [217 kms = 141 miles]

Guerrero Negro to Abreojos road. *The Highway crosses the flat Vizcaino Plain, paralleling the Cretaceous syncline, past dune fields and cultivated fields as it approaches tilted Miocene volcanic mesas. Cretaceous metamorphic and sedimentary hills form the skyline to the west. As the highway approaches the Abreojos road a series of steep-sided volcanic plugs are to the right.*

Abreojos Road to Tres Virgenes. *The highway turns east to cross the peninsula on the tilted surfaces of volcanic flows passing near numerous volcanic centers and dune fields. East of San Ignacio a strato volcano is approached and closely passed.*

Tres Virgenes to Santa Rosalia. *The highway drops down a series of fault controlled grades through volcanic rocks and then marine Pliocene rocks to the copper mining area of Santa Rosalia.*

217 To go straight to Santa Rosalia take the left fork at the triangle. If you have gone into town and are heading east from Guerrero Negro take the right fork at the triangle.

215 For many kilometers the highway will pass through a semi-stabilized dune field which consist of sand blown to the northeast from Scammon's Lagoon.

The vegetation growing on top of the dunes consists primarily of Mesquite (*Prosopis sp.*) and a few Saltbush (*Atriplex sp.*) plants.

MESQUITE is a large shrub or low tree which is found below 1500 meters in washes and on stabilized sand dunes throughout Baja. The pinnately compound leaves look like fern leaves. The small yellow flowers grow in slender spikes, the branches have pairs of nodal spines, and seeds are borne in tough pods that are eaten when they are ripe by rabbits, quail, deer, native rodents, cattle, and early Indians. Since Mesquite is one of the few trees which grows in the desert around Guerrero Negro, its sparse shade is welcome to a wide variety of terrestrial wildlife species and birds that nest in its branches.

208.1 The sign, "Parque Natural de la Ballena Gris", to the right side of the highway, marks the turnoff to Laguna Ojo de Liebre (Scammon's Lagoon) and a gray whale viewpoint 27 kilometers (17 miles) down a graded dirt road (the one perpendicular to the *Baja Highway*). The conditions of this road vary with the weather. (Check in town)

THAR SHE BLOWS! The best known of the great whales and the one most often seen along the Pacific Coast of Alta and Baja California is the California Gray Whale (*Eschrichtius gibbosus*). Gray whales feed in summer (mid-May to mid-October) in the plankton-rich western Bering Sea near Saint Lawrence Island and to the north into the Chukchi and Beaufort Seas. In winter (mid-October to early December) they migrate to the south along the more shallow Pacific coastal waters of North America to the various bays and lagoons of Baja California, especially Laguna Ojo de Liebre (Scammon's Lagoon), where they breed and give birth to their calves (early December through February). In the spring (March and April) they travel north along the coast and arrive at their arctic summer feeding grounds about the middle of May. Because they travel fairly close to shore it is relatively easy to see migrating Gray Whales from high points of land or from boats. It is estimated that the whales migrate about 10,200 kilometers (6,250 miles) each way every year by traveling 60 to 80 nautical miles (69 to 92 miles) per day at a speed of 4 to 4.8 knots per hour for 15-20 hours a day. As they travel, the frequently raise their heads out of the water "spy-hopping" to get their bearings. It is believed that they find their way on the long trip by memory and vision.

The California Gray Whale was once more numerous than it is today. In the 1800's Charles M. Scammon, a famous whaling captain and writer, estimated the gray whale population to be 30,000. However, Scammon and other whalers slaughtered the gray whales by the hundreds for their valuable oil. By 1937 it was estimated that their numbers had dropped to approximately 100. In 1938 they were given complete protection by an international treaty. Recently, counts of migrating whales from the shore and counts made in the lagoons of Baja from the air have shown that the gray whale population has increased to 15,000 to 17,000 and is still increasing yearly. Gray whales are mysticeti or baleen whales which feed mainly on small crustaceans (primarily a red shrimp-like organism known as lobster krill) (*See* 6:197) and plankton which are filtered from the water by sieve-like plates of "hair" in their mouths which is called baleen. The captured organisms are "licked" off the baleen by the whale's tongue and swallowed. Most feeding occurs during the Arctic summer months (mid-May to mid-October) but some occurs during migration and in Baja's lagoons.

THE REPRODUCTIVE CYCLE OF A FEMALE CALIFORNIA GRAY WHALE

June to October	24 hrs of daylight allow continuous summer feeding in Bering, Chukchi, and Beaufort Seas
November - December	Migration south, conception in route mid-Oct. to Early December
January - February	Wintering in Baja's lagoons
March - April - May	Migration north - March to mid-May
June to October	Summer feeding in northern seas mid-May to mid-October
November - December	Southward migration
January - February	Birth of calves and nursing (7 months)
March - April - May	Northward migration, nursing continues
June to October	Summer feeding in northern seas Nursing ends in August

Baleen whales are thought to have evolved in the warm temperate waters of the western South Pacific during the Oligocene. During the late Cenozoic they moved into the rest of the Pacific which included the northern seas presently inhabited by the California gray whale.

207 The stabilized sand dunes are vegetated with Datilillo, Mesquite, Jumping Cholla, Cardon, Palo Adan, Purple Bush, and Atriplex. The alkali flats between the dunes are only sparsely dotted with stunted annuals.

STABILIZED SAND DUNE

206.5 In early 1979 Pemex drilled an oil well approximately one kilometer east of the highway. In the future oil and natural gas may be another important source of income for residents of this region. There are a few scattered Datilillo, Cardon, Ice Plant, and Creosote Bush.

OIL AND GAS: The geosynclinal structure of the Vizcaino region is nearly identical to the great Valley of California. Large Geosynclines such as these produce much of the oil and gas in the world. Mapping in the Vizcaino and on the Magdalena Plain has indicated the presence of rocks and sedimentary environments favorable to oil and gas formation. After the traps where the petroleum accumulates have been delineated, this region may become a major oil producer.

202 This road metal quarry is dug in the Pleistocene limestones which mantle this area.

194 In this region of the Vizcaino Desert, the soil has become so dry that even the dunes are sparsely vegetated. The plants present along the highway are predominantly the xerophyte: Bursage.

189.5 This turnoff onto Mexico 18 leads to El Arco, a gold mining town and the site of a porphyry copper deposit in the metavolcanic rocks. The old road went inland at Rosarito north of Guerrero

Negro and skirted the hills to El Arco. It then headed south closer to Sierra San Francisco where it joins the highway near Abreojos Junction. Early travelers used the rougher road to avoid the sand near the coast.

Porphyry copper deposits are usually mined in bulk, on a large scale, and in open pits. The world's largest copper/molybdenite mine, which produces 90% of the world's copper supply, is at Climax, Colorado. The El Arco area promises to be one of the largest copper deposits in the world. They estimate 600 million tons of .7% copper ore.

187 The highly eroded volcanic mesas of the Sierra San Francisco lie ahead. To the right front are the volcanic plug domes of Sierra Santa Clara. To the right rear are the metamorphic and igneous rocks of Sierra El Placer. To the left and left rear are the metavolcanic hills near El Arco which are part of the southern end of the Peninsular Range Batholith. These mesas, domes, and ranges may not be visible at all on hazy days.

As you cross the Vizcaino Desert you will notice apparent changes in the vegetation. However, most of the species are everywhere, they just change in relative abundance. Datilillo, Cardon, Palo Adan, Creosote, and Pitaya Agria all put on a show for one reason or another, locally becoming the dominant species.

162 Jumping Cholla dominates the vegetation in this area. Plants commonly seen adjacent to the pavements edge along this stretch of the highway include several opportunistic species such as the tall yellow-flowered large-leafed roadside Indian Tree Tobacco, Cholla, and several species of grass.

A major **transpeninsular fault** roughly parallels the road in this area and provides a groundwater barrier which the farming enterprises are mining for irrigation. This fault is easily seen in the seismic and magnetic geophysical data that has been collected for this part of Baja.

To the east is the southern end of the Peninsular Ranges basement; it is largely covered by the Tertiary volcanic rocks of the Sierra San Francisco. There is very little Peninsular Ranges basement along the gulf coast south of this area.

The highway was built nearly parallel to the axis of a **Cretaceous syncline**, located a few miles to the west, with approximately 10,000+ meters of Jurassic to Paleocene sedimentary rocks. This geosyncline crosses the peninsula and heads southeast toward Loreto. The southern part of the peninsula to the cape region is entirely underlain by the Cretaceous syncline and overlain by Tertiary sedimentary and volcanic rock. The Sierra San Andreas (similar to the Coast Ranges of California) with the Franciscan assemblages and extensive exposures of sedimentary rocks is to the right in the distance.

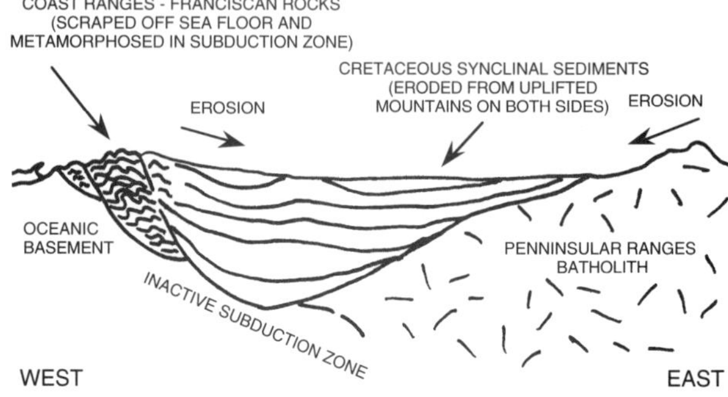

154.5 Ejido Francisco J. Mujica.
The massif of Sierra San Francisco to the left is a series of lava flows folded into a broad arch and tilted to the west. Some of the peaks rise above the main arch.

152 Before Km. 151 Creosote Bush, Palo Verde, Cardon, Garambullo, Pitaya Dulce, and Pitaya Agria are once again more abundant. After km. 151 Palo Adan and Datilillo along with many species of cactus become dominant.

144 The government agricultural cooperative and experiment station of Ejido Vizcaino is located at this junction. There is a gas station.

132 This turnoff leads to another government agricultural cooperative, Ramal Emilio Zapata.

124.3 Estacion Microondas de Los Angeles is to the right near a low volcanic hill.

WHERE ARE THE ANIMALS? Besides birds have you seen any other animals along the Highway? The most likely answer to this question is, "No". However, despite the harsh conditions of deserts (intense heat during the day, cold at night, and scarcity of water, vegetation, and cover) many kinds of animals do live successfully in this desert. Most of the larger desert animals are shy, and many are nocturnal and are unseen by daytime travelers. Those that are out during the day tend to be small in size and light in color for camouflage and protection against the heat and dehydration. Smaller desert mammals are mainly fossorial (fitted for digging and living in burrows). The lower temperatures and higher humidity of burrows help to reduce water loss by evaporation. The large desert ungulates (hoofed animals) such as pronghorn antelope cannot escape the desert heat by living in burrows. The glossy pallid color of their fur reflects direct sunlight, and the fur itself is an excellent insulation which helps to keep heat out. Additionally, heat is lost by convection and conduction from the underside of the antelope where the pelage (fur) is very thin. Ungulates conserve water by eliminating concentrated urine and dry feces.

SIDE BLOTCHED LIZARD

The Side Blotched Lizard is common in Baja. It eats scorpions, spiders, mites, and ticks. It loves to bask in the sun on rocks

If you want to see animals in the desert, the best time to look is at night. However, if you walk into the desert during the day, you may see many birds and small desert inhabitants such as colorful furry velvet ants (actually wingless wasps), lizards, black and red ants, and snakes. At the very least, you will see signs of many desert inhabitants such as the **tail drags** and **burrows** of rodents; **tracks** of birds, rodents, stink bugs, lizards, rabbits, coyote; and **scat** (feces) of animals. Several good field guides are available for identifying the tracks and scats of animals.

Some birds that you might expect to see in the Vizcaino Desert are:

COMMON NAME	LIKELY LOCATION
Amer. White Pelican	Flying over water
Cactus Wren	Among cacti
Calif. Brown Pelican	Flying over water
California Quail	Searching ground litter for food
California Thrasher	Searching ground litter for food
Gilded Woodpecker	Among cacti
Gray Thrasher	Looking in ground litter for food
Greater Roadrunner	Running on the ground
Ladder-back Woodpecker	Flitting through the air
LeConte's Thrasher	In sparse vegetation, flies when necessary
Red-Tailed Hawk	Soaring in skies
Sage Sparrow	Low perching in desert, sage brush or chaparral
Turkey Vultures	Feeding on carrion or soaring
Western Meadowlark	On fence posts or wires

118 This dirt road leads into the Sierra San Francisco.

THE SIERRA SAN FRANCISCO to the northeast is a mountain wilderness eroded out of a thick series of volcanic rocks. These rugged mountains rise from the surrounding desert plains to heights of more than 2,000 meters and stretch northward for 50 kilometers. The Sierra San Francisco range was once a rather simple basalt plateau roughly circular in outline. It is now folded into a broad westward-tilted arch crowned by Pico Santa Monica (2.144 m.). During the last few million years, erosion has produced the rugged landscape seen today. Hikers have compared valleys of the Sierra San Francisco to the topography of the Grand Canyon in Arizona.

Looking westward from their heights, sweeping views of Scammon's Lagoon and Vizcaino Plain can be seen. The peak of the volcanic Volcan las Tres Virgenes dominates the eastern horizon. Baja's best painted caves are found hidden in the canyons of these rugged mountains and in the Sierra San Pedro to the southeast. Mule trips to the caves are available in San Ignacio.

PICTOGRAPHS

PETROGLYPHS

PREHISTORIC CAVE PAINTINGS: The mountains of central and southern Baja California from the Sierra San Borja south into the Sierra San Francisco, San Pedro, and Sierra la Giganta are riddled with isolated caves painted by prehistoric people. These caves, located in some of the most inhospitable remote mountainous terrains of the peninsula, contain paintings of people, plants, unknown symbols, and animals. The figures are larger and more numerous than those of the famous prehistoric panels of Lascaux and Altamira.

Jesuit missionaries were probably the first "white men" to see Baja's painted caves. Leon Diguet, a French naturalist, was the first to publish an account of the caves after he visited them (1894). The caves were largely forgotten until the author, Earl Stanley Gardner, visited them in 1962 by helicopter. Since that time, many have visited the caves. An excellent book with beautiful photographs entitled *The Cave Paintings of Baja California* has been prepared by Harry Crosby.

117 The only plants that seem to be growing here are Datilillo, cactus, Jumping Cholla, and Cardon. Upon closer examination there are Acacia, many low herbaceous annuals, Mallows, Brittle-Bushes, and Sand Verbena.

112 The two flat-topped buttes ahead are composed of Miocene sedimentary rocks capped by dark basalt. They are erosional remnants of the volcanic mesas of the Sierra San Francisco.

108 The highway leaves the very flat old lagoon surface and begins to climb the gently-sloping alluvial fans which are cut by washes.

106 Common Ravens are often seen in this area.

COMMON RAVEN OR AMERICAN CROW?

There are three easy ways to distinguish between these two black birds: tail shape, habitat preference, and flight patterns. Both birds are black and may be a variety of sizes depending on their age. Color and body size are not good criteria to use. The chart will help you identify these birds:

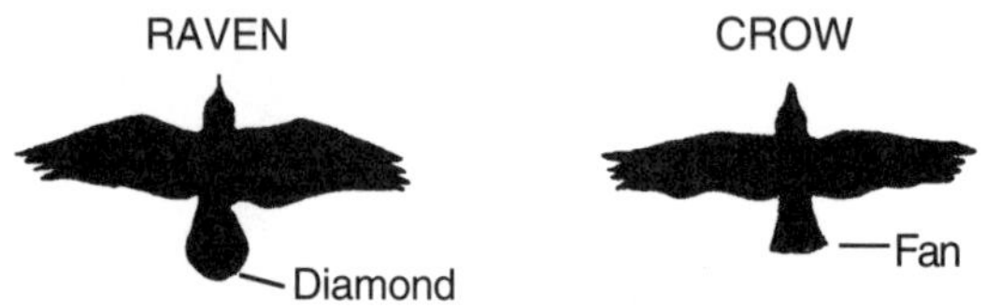

	Raven	Crow
Tail shape	Diamond	Fan
Habitat Preference	Deserts & cities throughout the peninsula	Not usually seen south of San Quintin
Flight :	Can glide for more than 3 seconds	Have a short gliding time - less than 3 sec.

102 Some of the Elephant Trees, Palo Adan, and Datilillo which grow here support the harmless epiphytic ball moss, *Tillandsia*.

98.5 The Pleistocene limestone, which covers parts of the middle of the Vizcaino Plain, is exposed along the highway. This limestone was deposited in a shallow warm sea which covered this area within the last million years.

The lagoons of San Ignacio and Scammon are small remnants of this embayment which probably isolated the Sierra San Andreas and the Santa Clara Buttes as islands.

98 At the turnoff to Punta Abreojos the gray Miocene marine sedimentary rocks of the San Ignacio Formation are exposed under the basalt on the high hill to the right. Just beyond the turnoff are more exposures of the Miocene sedimentary rocks. To the right are the Santa Clara Buttes. The Santa Clara Buttes are Miocene (10-15 million years) andesite plugs which have intruded into and domed the Cretaceous and Paleocene sedimentary rocks. Their temporal relationships and geology are identical to the Marysville Buttes area of the Sacramento Valley in California.

SANTA CLARA BUTTES

The isolation of the Vizcaino: The road to the right leads approximately 90 kilometers to the west to an isolated section of the Pacific coast and the small cannery and fishing settlement of Abreojos. It is generally a fairly good graded road which most two-wheel drive vehicles can make. The area around Punta Abreojos is one of the favorite surfing spots on the peninsula. The break of the waves off the point enables some surfers to have runs up to a kilometer long. The small isolated lagoon at Rene's Fish Camp offers excellent dry camping and fishing

97 Southeast of this point the vegetation is dense with Datilillo, Cardon, Jumping Cholla, Pitaya Dulce, Cardon, Palo Blanco, Elephant Trees, occasional Lomboy, Leatherplant, Palo Verde, Acacia, and occasional Ball Moss.

91 There is a large dip and bend in the highway. The flat-topped mesas ahead on the right are the mesas of the Sierra San Pedro.

The lagoon of San Ignacio with the two white Salitrales (salt pans) on either side is in view to the right.

88.2 This road leads to Microondas Albun microwave tower. The view from this tower is sweeping. On a clear day you can see south to Sierra San Pedro and the mesas of Cuarenta, San Ignacio Lagoon, the Sierra Santa Clara plugs, and the mesas of Santa Clara to the west. To the northeast is the mountain mass of Sierra San Francisco.

The road quarry at the foot of the road to the microwave tower is in the San Ignacio Formation and has yielded some spectacular agatized (a translucent cryptocrystalline variety of white quartz) Turritella (*Turritella sp.*) fossils which are just below a light-colored volcanic tuff (ash) bed.

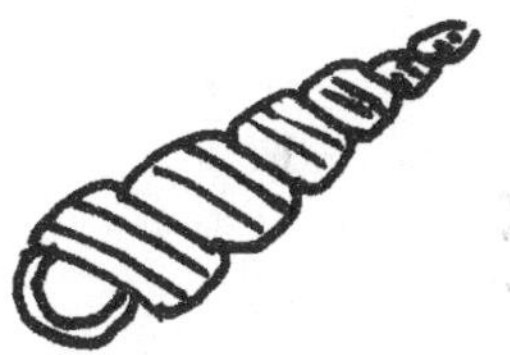

AGATIZED Turritellas: Approximately 11 million years ago this area was covered by a shallow sea, and numerous molluscs including *Turritella ocoyana* flourished in that sea. A volcanic eruption spewed ash over the area; this resulted in a mass mortality of these molluscs. Over the years groundwater leached the calcite from the fossil shells and replaced it with silica from the volcanic ash and produced agatized Turritellas.

86 The highway climbs and eventually skirts south of the foothills of Sierra San Francisco.

The road follows what amounts to a stripped-dip slope developed on a resistant bedding surface with basalt hills along both sides of the highway.

84 To the rear (southwest) is a good view of San Ignacio Lagoon (Salitrales). On a clear day you can see across the Vizcaino Desert to the peaks and mesas of Sierra Santa Clara.

82 The highway dips into an arroyo cut into the lava surface.

81 The small red-brown conical hill to the left is an extrusive volcanic plug.

78.5 Several dark conical-shaped cinder cones dot the landscape. The very large dark mountain in the far distance ahead is Volcan las Tres Virgenes.

78.25 This is the turnoff to a paved jet airport (Areopista) which serves San Ignacio.

76 The road passes along the edge of several dark basalt mesas with views of the date palm oases of San Ignacio and San Lino. These mesas are capped by dark basalts and underlain by the white and greenish San Ignacio Formation.

Mina (1957) defined the **San Ignacio Formation** for the light-gray sandstones, tuffaceous sandstones, and tuff outcropping near the town of San Ignacio. The San Ignacio Formation is a marine facies of the Comondu Formation which overlies and interfingers with the Isidro Formation. The San Ignacio Formation is Middle Miocene in age in the San Ignacio area; a volcanic tuff bed was K/Ar dated at 11 m.y. Some beds contain the agatized Turritellas (*Turritella ocoyana* and *Turritella inezana*). Further south part of the unit has been K/Ar dated at 22 m.y.

74.5 Excellent exposures of the San Ignacio Formation are seen here. The conspicuous white building in the middle of the palm oasis ahead is the mission church.

74 This is the turnoff to San Lino. This small community is a "suburb" of San Ignacio.

73.5 This side road leads 3 kilometers to San Ignacio. Ground water surfaces in the arroyo above town forming a flowing river with an oasis of imported European date palms. It's well worth a turnoff into the town to see the square which is shaded by large Indian Laurel Fig trees and the impressive stone Mision San Ignacio de Kadakaman built by the Dominicans in 1786. Hotel La Pinta de San Ignacio is located on the road into town.

MISION SAN IGNACIO DE KADAKMAN

SAN IGNACIO RIVER

HUERTAS DE SAN IGNACIO - Huertas is the Spanish name for garden or orchard, and it might be aptly applied to the desert oasis which surround the two small communities of San Ignacio and San Lino. It is reported that over 80,000 date palms grow here. Date Palms (*Phoenix dactylifera*) were introduced for cultivation into mainland Mexico by Jesuit priests and have subsequently been cultivated in San Ignacio (since 1728), Mulege, Loreto, and San Jose del Cabo.

73 The date of 11 m.y. was obtained on the San Ignacio Formation in this roadcut.

71 The tuffaceous sandstones of the Comondu Formation form the brown outcrops along the edges of the arroyo.

69 The basalt at this location was dated at 9.7 my. The highway begins an ascent of a grade bordered by a dense cardonal. The predominant plants in this area are Cardon, Cholla, Elephant Trees, Palo Verde, and Pitaya Agria. This area has been overgrazed by domestic animals which allows the cacti to dominate the vegetation. The old road is visible in the canyon bottom to the left. This was a short, very rough grade for earlier Baja travelers.

67 This is the first good view of the composite strato volcano, Volcan las Tres Virgenes. Tres Virgines is composed of a combination of clinkery basalts and andesites and pyroclastic material in the form of Lahars and associated breccias and mudflows.

TRES VIRGINES VOLCANO

Volcanoes are classified on the basis of the type of cone which they produce, and the type of cone depends on the composition and viscosity of their lavas. Basaltic lavas with a lower content of silica (50%) flow easily, while rhyolitic lavas which have a higher content of silica (70%) resist flowing (are more highly viscus), and are more explosive. Basaltic lavas from cinder cones have a higher content of gas and are also more explosive. As the gases are released from lavas the basalt becomes more fluid and forms shield cones and lava plateaus.

The world's well known volcanoes are **Strato Cones**; examples are Mount Vesuvius and Krakatoa. They are usually 2-3 miles high, tend to dominate the landscape, and are the most destructive types of volcanoes.

The table relates the types of cones with activity, rock types formed, and material ejected.

TYPES	*ROCK TYPE*	*CONE*	*MATERIAL*
EXPLOSIVE	BASALT	CINDER	CINDERS+
	RHYOLITE		FEW FLOWS
INTERMEDIATE	ANDESITE	STRATO	TUFF & LAVA
QUIET	BASALT	SHIELD	LAVA FLOWS+
			FEW CINDERS
FISSURE	BASALT	PLATEAU	THIN LAVA
			FLOWS

65 The small brownish hill ahead with several sharp peaks is Cerro Colorado, a sub-recent basaltic eruptive center.

62 Between here and Kilometer 61 fossil dune deposits which exhibit cross-bedding are exposed in the roadcuts.

59.5 This is the road into Santa Marta and the southern Sierra San Francisco.

58 This long straight stretch of road presents views of the volcanic mesas and plains of the middle part of the peninsula. The massif (French for massive) of Sierra San Pedro is on the right, and the relatively unexplored rugged volcanics of Sierra San Francisco is on the left. The large dark mountain straight ahead which dominates the horizon is Volcan las Tres Virgenes.

VOLCAN LAS TRES VIRGENES: This composite volcano is thought to have erupted as recently as the early half of the 18th century. There is a report from one of the early Spanish missionaries of an eruption in this area in 1746 which may well have been a description of the most recent eruption of Volcan las Tres Virgenes. There is another report of some smoke in 1857.

The activity of volcanoes is not readily apparent to us in our short life spans.

An **ACTIVE** volcano has a record of historic activity such as Mount St. Helens in the state of Washington. The lack of active volcanoes in Baja may be due to the short recorded history of Baja or due to the fact that the sliding action of plates does not tend to produce volcanic action.

A **DORMANT** volcano does not have a record of historic activity. It does have good relatively unerroded shape and shows signs of sub-Recent activity. Most of the Cascade Volcanoes such as Mount Rainier are considered dormant. Las Tres Virgenes appears to be a dormant volcano.

An **INACTIVE** volcano shows no sign of activity and shows significant cone erosion such as Crater Lake in the state of Oregon.

Magmas are molten or liquid rocks that originate at the base of the earth's crust and in the upper mantle. The majority of the active volcanoes today are in the Subduction Zones around the Pacific Ring of Fire and in the Alpine-Himalaya Belt. The **friction** generated in the Subduction Zones causes the melting. Another belt of active volcanoes is on the Mid-Ocean Ridges where the **release of pressure** causes the melting. The underlying source of the latent heat is the **radioactive decay** of elements. Volcan las Tres Virgenes relates to the formation of the Gulf of California along the East Pacific Rise.

54.5 Signed road to Cave Paintings.

51 There is an abundant supply of loose sand in this valley. The wind in this area and the supply of sand have combined to form the dunes which cover the faces of many of the hills.

44 Some of the tallest and most beautiful Elephant Trees (*Pachycormus discolor and Bursera microphylla*) of the peninsula grow on the lava flow to the southeast.

The vegetation on the flat plain (Llano) between and around the lava flows consists of Brittle-Bush, Mallow, Palo Verde, Creosote Bush, Leatherplant, Pitaya Dulce, Cardon, Teddy Bear Cholla, Hedgehog Cholla, Garambullo, Jumping Cholla, Palo Blanco, Palo Adan, and Lomboy.

ELEPHANT TREE

ELEPHANT TREE IN BLOOM

40.6 Just before the first lava flow is one of the few places on the peninsula where two types of Elephant Trees (*Pachycormus discolor* and *Bursera microphylla*) occur together so close to the road. A walk onto the flow leads into a jagged landscape with Elephant Trees. This is also a good place for a close-up view of the floral dominants of Llanos de Magdalena area introduced at Km 73.

Elephant Trees - are represented by the unrelated species of the two genera *Pachycormus* and *Bursera*. They grow luxuriously as trees which store water in the cortical cells of their elephantine trunks. The *Bursera* are easily distinguished from the *Pachycormus* by the incense-like odor of crushed leaves or fruits. (*See* 4:9)

LOMBOY

Lomboy - is a member of the spurge family commonly seen on the flats, bajadas, and mesas of this region. This plant produces an acrid sap and remains leafless most of the year. This is an adaptation which reduces water loss by transpiration in the extremely arid environment of the Magdalena Plain desert. Leaves appear after each infrequent rain and turn a distinctive red before falling off as the soil dries out again.

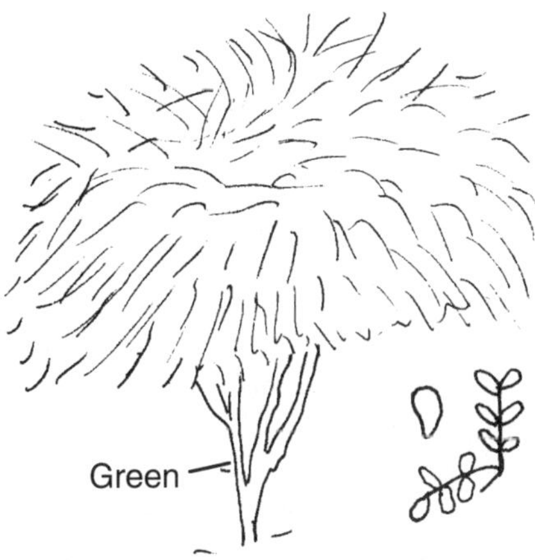

Palo Verde - is a green barked tree which produces legume-like "bean" pods. However, Palo Verde is a member of the Senna family. Like Lomboy, this tree is normally leafless. The green trunk is a cladode, a stem that acts like a leaf which performs photosynthesis in the absence of leaves. (*See* 9:210)

Palo Adan - is a relative of the Ocotillo. However, Palo Adan has thicker branches, a trunk, smaller flowers, and normally grows from Parallel 28° south to the Cape Region in the clay and granitic soils of alluvial plains. Ocotillo is not usually found below the 28th parallel and is replaced by its look-alike relative, Palo Adan (Adam's stick). (*See* 4:16.5)

Creosote Bush is a very common resinous desert shrub of the southwestern U.S. and Baja. Unlike many other desert plants, it is green the year round. The waxy coating on its leaves reduces water loss by transpiration which allows the shrub to withstand long periods of extreme drought. The strong-smelling resinous leaves resemble the odor of the coal tar distillate-creosote. Notice that very few plants grow under Creosote Bush. The roots of this shrub produce a toxic substance that inhibits the growth of other plants except after heavy or frequent rains when it's leached from the soil which permits the growth of desert annuals. As the soil dries, the inhibitor accumulates and poisons the outsiders. This phenomenon is known as allelopathy and helps eliminate competition for water in a dry desert environment. Creosote Bush was considered a cure-all by Baja's early Indians, and early Spaniards used a decoction of this shrub to treat sick cattle and saddle galled horses.

Other plants associated with the plants listed above are Jumping Cholla, Cardon, and pavement edge opportunists such as Brittle-Bush and Indian Tree Tobacco.

TRES VIRGINES

40 The highway passes close to the west side of Volcan las Tres Virgenes and its multiple lava flows. The ground for miles around is covered with pumice of an unknown origin which has been shown to be unrelated to the activities of Volcan las Tres Virgenes.

39.2 Rancho El Mesquetal on the right.

38.7 The highway parallels the edge of a blocky clinkery basalt lava flow from Volcan las Tres Virgenes. The deep clefts (cracks) on the flow were caused by the pressure of the molten interior which pushed on the solid crust much like the cracking on the crust of a loaf of bread. A number of unusually large Elephant Trees are growing on this lava flow.

38 The highway begins to climb through a small canyon on the western slope to the crest of the range and the Virgenes grade.

35.1 This is the top of the Virgenes' grade where the highway begins a long descent into the Gulf coast drainage southeast of Tres Virgenes. Descending this grade used to be extremely difficult and dangerous.

TRES VIRGINES GRADE

There was a shrine at the top of the grade and another at the bottom along the old dirt road. Travelers would stop at the top to pray that they would make it down and again at the bottom to give thanks that they had. The very difficult switchbacks of the old road are visible over the scarp just beyond the crest of the grade.

CHANGING VEGETATION REGIONS The highway descends into the eastern drainage of the Sierra San Pedro, into the Gulf Coast Desert area of the Desert Region. The Gulf Coast Desert area extends south from Bahia de Los Angeles along the eastern Gulf coast of the peninsula to just south of La Paz; it includes most of the southern Gulf islands. The predominant plants of the Gulf Coast Desert are Cirio, Ironwood, Palo Adan, Creosote Bush, Brittle-Bush and three species of Elephant Trees (*Pachycormus discolor*, *Bursera microphylla*, and *Bursera hindsiana*).

33.8 There is a fault in this roadcut.

32.3 The highway crosses an eroded fault scarp which forms the base of the grade. This fault has a major lateral motion component. This fault and one to the north of Tres Virgenes may be causing a small rift area which is responsible for Tres Virgenes and two other volcanoes on the area. The hill to the left, Punta Arena, is a volcanic eruptive center which has domed the strata of Mesa El Yaqui into a hill along part of the Gulf Fault Zone.

32 The road to the left leads to a large geothermal project (Proyecto Geothermico Las 3 Virgenes) located a few kilometers northward. This project is under the control of the Commision Federal de Electricidad and will produce electricity by harnessing geothermal heat to heat water to steam which in turn drive a steam turbine generator to produce electricity for this part of Baja. They estimate a production of 15 M W.

31 Rancho Las Virgenes. There is a view to the rear of Volcan Tres Virgines.

TRES VIRGINES FROM THE EAST SIDE

30 The highway follows a relatively narrow surface developed on the flat-lying volcanic and fluviatile sedimentary rocks of the Comondu Formation.

18.5 This is the first view of the Sea of Cortez (on a clear day). The volcanic cone visible offshore is the gulf island of Isla Tortugas. There is a good viewpoint, a short hike to the right, near the top of the grade.

17.5 The highway begins a twisting descent of the infamous Cuesta (grade) del Infierno in the Barranca Palmas (Palm Canyon) through yellow Pliocene marine sedimentary rocks. It is easy to imagine this descent on the old narrow rutted dirt road, and to understand why it is infamous to those who made that descent before the road was widened, leveled, and paved. This section of the highway is very steep and has few turnouts or shoulders. On our last photo trip we were taking a photo at the first turnout when a large truck rolled over on the curve below us.

OVERTURNED TRUCK ON GRADE

THE THREE FORMATIONS OF THE SANTA ROSALIA PLIOCENE: The Pliocene sedimentary rocks of the Santa Rosalia area were deposited near a shoreline in a shallow marine environment interrupted by the deposition of conglomerates and tuffaceous material derived from explosive volcanic eruptions. It is part marine and part nonmarine and contains gypsum deposits.

During the Pliocene the fault along the Gulf was active and periodically uplifted the area west of Santa Rosalia. This uplift resulted in erosion which initiated the deposition of conglomerates into the finer-grained sandstones and siltstones.

Mineralizing solutions which rose along the faults flowed through the more porous and permeable conglomerates and mineralized the clays into the Boleo copper deposits.

PLIOCENE SEDIMENTS FROM VIEWPOINT

The Pliocene of the Santa Rosalia area has been divided into three formations separated by unconformities and characterized by faunas which are believed to be Early, Middle and Late Pliocene (Wilson, 1948, Wilson & Rocha, 1957):

The Lower Pliocene **Boleo Formation** is a succession of sandstones, tuffs and conglomerates which contain copper and manganese deposits. Recently the Cinta Colorado Tuff has been age dated at 6.8 m.y. This places this unit in the uppermost Miocene. The Middle Pliocene **Gloria Formation,** later renamed the Tirabuzon Formation, is a sequence of fossiliferous marine sandstone, siltstone, and conglomerate. It can be seen between Km 4 and 8 north of Santa Rosalia. The Upper Pliocene **Infierno Formation** is a succession of fossiliferous marine sandstones and conglomerates.

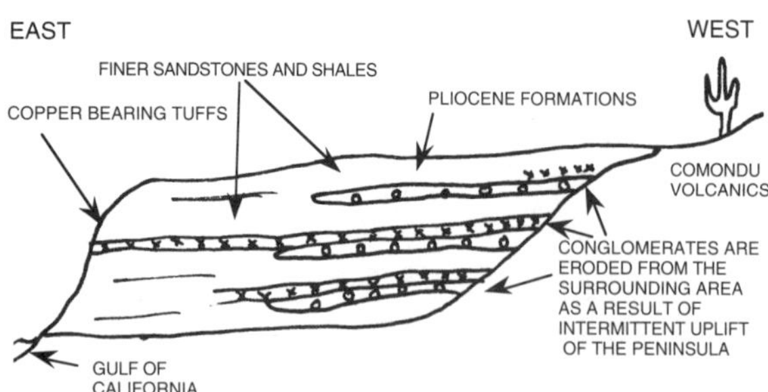

Boleos in the Desert: Jose Villavivencio set out on horseback in 1868 to find a route across the barren mesas to the village of Santa Maria. On the way his horse dislodged a rounded bright blue and green ball of mineralized copper carbonate rock known as a boleo. This discovery led to the formation of the Santa Rosalia copper district under the operation of the French Compagnie Boleo. It was ranked as the second largest copper producing district in Mexico with over 500 kilometers of underground workings. Villavivencio was paid 16 pesos for his boleo discovery.

In 1885 Compagnie Boleo obtained from the Mexican government all the claims for the 200 square kilometer Boleo district. It developed the ore bodies and built the town of Santa Rosalia, mining camps, the harbor, and a smelter which utilized blast furnaces. Copper matte was shipped to Guaymas, then by rail to New Orleans, and finally as ballast in steamers to France.

In 1922 the blast furnaces were replaced by a more efficient reverb-converter combination. This produced a blister copper product which was shipped to Tacoma, WA, in company steamers which returned to Santa Rosalia with fuel, petroleum coke, lumber, and mine timbers.

The area declined after 1930 as the high-grade ore was mined out. However, World War II and some government assistance kept Compagnie Boleo in operation until its final demise in 1950.

Atypical Ore Deposits: The Boleo district deposits are thin gently-dipping tabular bodies of relatively impervious clayey tuffs which overlie sandstone conglomerates. The main gangue mineral is a montmorillinite or bentonite clay, which is not typical of copper deposits. Low temperature hydrothermal solutions appear to have ascended through cracks and faults in the volcanics. They were blocked by the impermeable tuffs and permeated the tuffs by diffusion. This left the copper minerals finely dispersed in the clay matrix.

Underground mining developed by Compagnie Boleo consisted of driving entries to intersect the beds from the deep arroyos. Irregular drifts and room-and-pillar stopes and later more systematic methods featuring conveyor belts at the faces and double-tracked, mule-powered entries were used to mine the thin ore beds. The soft deposits did not need explosives, but extensive timbering was required. An overhead tram was used in one mine, and a network of narrow-gauge rail lines with Baldwin locomotives connected the mines with the smelter.

WORKING COPPER MINE

RAILROAD RUNNING STOCK

Smelting: Boleo ores are sulfide deficient for the purpose of making matte in a reverb. Local gypsum is used as a source of sulfur, and coal or petroleum coke is added to reduce gypsum sulfate to sulfide for matte production. The calcium in gypsum serves as a flux for the predominantly siliceous Boleo ores. The 25-30% moisture content of the bentonite gangue requires pre-reverb drying in a rotary kiln. The reverb is choke fed by a charging system which employs drag conveyors on top at the side walls. Reverb slag is stacked or disposed of at sea, and 99.5% pure blister is produced from an eight hour blow of the matte. Siliceous material from the Mulege area south of Santa Rosalia is also added as flux.

Present operations at Santa Rosalia are admittedly uneconomic but are sustained to support the local community while attempting to devise a viable and economic metallurgical future. Ore of the required minimum smelter feed grade is supplied by small-scale hand mining operations in two underground mines and in several small surface pits known as "descapotes".

15.9 Numerous faults are visible in exposures of light yellow-brown Pliocene sedimentary rocks.

13.5 The highway crosses a major arroyo where, for the first time, Palo Blanco trees are seen along the highway. Elephant Trees, Cardon, Creosote Bush and stands of Palo Verde are also visible.

PALO BLANCO

13 Nothing grows on this badland's topography. Due to the high percentage of gypsum in the soil, this region is considered to be a saline physiological desert (*See* 16:35.5).

11.75 This roadcut exposes massive well-bedded gypsum layers of the Pliocene. There are massive Gypsum crystals in some arroyos.

10.5 To the south Isla San Marcos and its gypsum beds is now visible (*See* 6:188).

7.3 The highway descends to the south along the Gulf shore toward Santa Rosalia (Cachanla). To the left is dirt road which heads north to an airstrip and a camping beach called Playa Santa Maria. On one trip I almost drove off the road as I came around this corner and found myself looking a pilot in the face as he was landing.

GYPSUM CRYSTAL

The shiny road base of this section of the highway is from the smelter tailings of the mining operation. The playa is composed of small cobbles and shingles of rock.

4 Between here and Kilometer 3.5 yellow Pliocene beds of the Tirabuzon Formation are well exposed in the cliffs and roadcuts.

3 A view of Tres Virgenes is seen to the rear. The dark hills to the right between here and town are parts of the tailings dump for the Boleo Copper District and the Lucifer Manganese district.

0 To enter Santa Rosalia turn right. Directly to the front is a steam engine monument which is a reminder that this town grew to house a population in excess of 12,000 inhabitants in its hey day. The French-owned mining company, Compagnie Boleo, began mining operations in the 1880's and ended in the mid 1950's as the quantity and quality of ore decreased. The rustic French architecture and Eiffel's metal church are worth seeing.

Its worth turning right in front of the church and going up the hill to the mining museum, old train engines and rolling stock, and the Hotel Francis.

BALDWIN LOCOMOTIVE

CHURCH BY EIFFEL
AN EIFFEL THAT IS NOT IN PARIS

The sheet metal church in Santa Rosalia designed by A. Gustav Eiffel, the architect of the Eiffel tower in Paris, was brought to Santa Rosalia from France. The prefabricated metal church was originally displayed at the 1898 World's Fair in Paris, France. It was stored for many years, purchased by Compagnie Boleo and brought to Santa Rosalia. Look inside for a view of Eiffel's structural genius.

CAUTION! Most of the streets in Santa Rosalia are one-way. There is good shopping, and most of the stores, supermarkets, panaderia, and pharmacy are on the main road into town. The meat market is one block south of the supermarket.

Log 6 - Santa Rosalia to Loreto [196 kms = 122 miles]

Santa Rosalia to Mulege*. The Highway follows alluvial fans along the narrow fault controlled gulf coastal plain fringed by Miocene volcanic mesas. Near Mulege it climbs the Miocene volcanic hills to descend into Mulege with its Pleistocene terraces.*

Mulege to Loreto*. The Highway continues south along a major fault zone on the same coastal plains to Conception Bay with the Miocene volcanic mesas close on the right. The opposite side of the bay consists of faulted Miocene volcanic rocks. South of Conception Bay the road climbs through a pass in the volcanics and follows a major alluviated graben along the Gulf fault zone with the same volcanic mesas to the west and faulted volcanic rocks to the east. Near Loreto it passes through a serles of faulted marine Pliocene sedimentary rocks.*

197 In 1997 the ferry terminal at Santa Rosalia had regularly scheduled service to Guaymas.

On several occasions the author of this guide has seen innumerable beached pelagic red crabs (Krill) covering this beach in a layer almost a half meter thick.

KRILL ON BEACH AND HARBOR

Krill: Scientifically, biologists place these beautiful diminutive crustaceans in the Order Euphausiacea. Euphausiids are small pelagic marine shrimp like forms which differ from decapods in having biramous thoracic appendages and in other characters; most are luminescent, and all have well-developed pleopods for swimming. Euphausiids or "krill", important as food of whales, penguins, and several species of fish, are generally deep-living, but many perform diurunal vertical migrations and rise to the surface at night. Locally, schools may be encountered swimming near shore, and occasionally species of *Thysanoessa* and *Euphausia* may be washed ashore.

KRILL ON BEACH

They were often observed by early travelers. The Dominican, Louis de Sales wrote in the late 1760's that "while the ship Venus was thus following the coast to enter the Gulf of Cortez, the lookouts noticed red patches of whales blood. The presence of the whale-shops within sight might justify this guesswork supposition which however, was soon contradicted by the reality for soon, as we advanced, we crossed over these patches and recognized that their red color was caused by a multitude of small, vermilion colored crustaceans. These crustacea were [like] big shrimps, but they had what shrimps lack, pincers like those of lobsters." The Gulf of California is referred to as the "**Vermillion Sea.**"

For the next two kilometers the road passes through numerous exposures of the Pliocene Tirabuzon and Boleo Formations and occasional pink tuffs of the Comondu Formation.

193.8 The beds ahead are gypsum and tuffaceous sandstones of the Pliocene Boleo Formation.

192 The highway crosses Arroyo de Santa Agueda and the Santa Agueda Fault and passes through exposures of both the Boleo Formation and the Infierno Formation to climb a grade through sandstones of the Pliocene Infierno Formation.

The head frame of the abandoned San Luciano copper mine can be seen below in the canyon to the right. The remnants of the town site of San Luciano (streets, foundations, etc.) are still visible on the gentle slopes behind the mining equipment. The hills around the mining region of Santa Rosalia are riddled with mines, shafts, and the remains of mining structures and equipment.

188.5 This Ramal (side road) leads to Santa Agueda which supplies water for Santa Rosalia.

188 For the next several miles there are good views of Isla San Marcos.

ISLA SAN MARCOS is a Miocene volcanic island covered with Pliocene marine sedimentary rocks which are primarily composed of gypsum precipitates deposited during the Pliocene in shallow marine basins with restricted circulation in the Gulf of California. The barren whitish slopes near the south end of the island are the gypsum deposits in the Pliocene Marquer Formation. The gypsum is mined as a retarder in Portland cement, a flux in copper smelting at Santa Rosalia, and in making Plaster of Paris. At the tip of the south end of the island is the village of San Marcos where miners and fishermen live. One of the best views of Isla San Marcos is at Kilometer 176. Some people see the cartoon dog "Snoopy" lying in wait when they look at the outline of Isla San Marcos.

The vegetation is Palo Verde, Elephant Tree, Cacti, Cardon, Cholla, Leatherplant, Lomboy, and many desert annuals.

184 There is a good view of Santa Agueda and the palm shaded cove of San Lucas.

SANDY SPITS AND BEACHES: Most of the features along Baja's shoreline are produced by erosion or deposition. Depositional features along the Gulf shore in this area are built of eroded and weathered materials brought down to the shore by streams from the eastern escarpment of the Sierra la Giganta or material eroded from headlands by waves. At Santa Agueda sand/mud has been transported south by the waves. This depositional feature forms the sand/mud spit of Santa Agueda which is a fingerlike extension of the beach. Other materials have been moved by currents and deposited to form the sandy/mud beach of San Lucas cove.

179.8 This side road leads up through the Comondu Volcanics to the Microondas San Lucas. There are many picturesque specimens of Elephant Trees near the top. There is a tremendous view of Volcan Tres Virgenes, and the Santa Rosalia area to the northwest, and Bahia de la Concepcion in the southeast. The road is paved to the top with cobblestones.

174 Best view of Isla San Marcos

169 This side road leads to the village of San Jose de Magdalena. The village, located in a palm oasis, originated as a Jesuit visiting station of Mision Santa Rosalia de Mulege. Today it is a agricultural community which grows vegetables, citrus, and dates. The road to the village is rough but scenic. Ugarte built a 500 ton ship for exploring the Gulf from oak timbers cut in the area.

GULF DESERT FLORA AND PALM OASIS

163 The view to the southwest is of the high Gulf escarpment which is not very steep in this area. Between Santa Rosalia and Mulege the Gulf Fault Zone is a broad series of faults which cause the rugged foothills to the west of the highway.

159 The flora along this section of the highway is characteristic of the Gulf Coast Desert and is dominated by Cardon, Palo Blanco, Desert Mallow (apricot flowered), and several species of cacti and yellow-flowered annual composites. The annual composites and Desert Mallow are primarily growing opportunistically along the pavement edge or in the disturbed soils which border the highway.

156 The graded road to the left leads to a resort 20 kms. from the highway at Punta Chivato. Exposures of fossiliferous marine Pliocene beds can be seen in the sea cliffs on Punta Chivato.

145 The highway follows an alluviated plain before it climbs a grade at Km. 143 through the volcanic rocks of the Comondu Formation. This plain drains southwest into a gap in the hills to join Rio Santa Rosalia in Mulege.

139.5 Ahead and to the left is a multicolored hypabyssal plug which is one of several 20 million year old intrusive plugs exposed in this area.

138.3 This is the first view of the desert palm oasis village of Mulege. The highway begins a descent through the volcanic strata.

Mulege is a refreshingly beautiful spot along the banks of the Rio Santa Rosalia de Mulege (locally known as "Rio Mulege"). Mulege is located in a valley bottom whose sides resemble an extremely dry desert cactus garden. Like San Ignacio, Mulege began as a Jesuit Mision and was called Santa Rosalia de Mulege in 1705. The name was later shortened to Mulege, as the northern French operated mining town of Santa Rosalia took the name "Santa Rosalia" due to its larger size and notoriety. Because of the presence of the river, a oasis of European date palms, citrus, olives, grapes, and oranges has developed here. Slightly upstream from Mulege (inland), the waters of the Rio Santa Rosalia have been dammed and diverted to irrigate palms, mangos, papayas, bananas, and citrus.

135 Purple andesite breccias are located at the north end of the Puente Mulege. The road to the left leads into Mulege along the north side of the Rio Santa Rosalia.

A CALL TO ARMS! One of the many battles of the Mexican war was fought in Mulege in 1846. Go through town and down the north side of the estuary to El Sombrerito. Climb on the 20 million year old intrusive mass of El Sombrerito and imagine that you are commanding the American forces which have just landed from the sloop, U.S.S. Dale. The Mexicans hold the hill to the north across the sand spit. Needless to say, the Americans soon withdrew from their indefensible position and Mulege was "saved".

GREAT BLUE HERON ON MANGROVE

MULEGE ESTUARY AND EGRET

On the bridge there is a view of the Rio Santa Rosalia de Mulege, the estuary, and the now closed federal prison on the hill to the north (large white mission-like structure). In the past the prisoners were let out of the jail to work during the day. At the end of the day, the prison bell would ring, and the prisoners would return to their cells. Few of them tried to escape, since recapture meant a trip to Islas Tres Marias and its escape-proof prison.

The conical hill behind the prison is the same Miocene intrusive volcanic plug, seen as the highway entered from the north of town.

133.6 The roadcuts here contain altered gypsiferous Pliocene strata cut by basalt dikes. Most of the yellow sandstone and siltstone layers in this part of Baja are probably Pliocene.

133.5 The lighthouse to the left is where the battle of Mulege was fought.

EL SOMBRERITO

130.9 Just before the entrance to La Serenidad Lodge, the roadcut exposes yellow-brown sedimentary rocks of Pliocene age which contain oyster fossils. Overlying them are fossiliferous Pleistocene deposits of the Mulege Terrace. The Mulege Terrace is a 10 meter marine terrace of Late Pleistocene age which is extensively exposed in and near the town of Mulege.

FOSSILIFEROUS MULEGE TERRACE SEDIMENTS

A STABLE SHORELINE: In this area the Gulf shoreline has remained relatively stable since the Pliocene. There are nearshore Pliocene rocks exposed along the Mulege Estuary and nearby coastal areas. These Pliocene rocks are overlain by the Pleistocene Mulege Terrace with its diverse shoreline fauna. The present shoreline is only a few meters lower and nearby.

THE MULEGE TERRACE has been dated at 125,000 to 145,000 years (Ashby, Ku, & Minch, 1987) and probably represents the Sagamonian 5e high stand of sea level. The fauna contained in the terrace deposits indicate a variety of environments including: open rocky shoreline, protected sandy shoreline, and estuarine (Ashby & Minch, 1987)

126 To the left is a good view of Punto Gallito with its Pleistocene marine terrace and dune field. The north end of Punta Concepcion comes into view.

BAHIA DE CONCEPCION: Twenty five miles long, it is one of the most beautiful bays on the Gulf coast. Turquoise and aquamarine waters wash its steep rugged shores with the milky blueness of tropical seas the world over. The color of the waters, produced by colloidal limes derived from the shells and skeletons of a myriad of marine animals, is the same as the color produced by colloidal suspensions in glacial streams and travertine springs, such as Havasupai in Arizona. Biological, "shell-sand" beaches are found in sheltered coves, such as Santispac, Coyote, and El Requeson. Many of the coves are bordered by small mangrove swamps. Mountains covered with desert vegetation almost surround Bahia de Concepcion and make it a land of extreme contrasts between mountain, desert, beaches, and tropical blue water. At the south end of the bay, is one of the densest cardonals in Baja.

124.5 A cinder blanket is visible in this roadcut. The road to Microondas Tiburones circles around an ancient cinder cone. From the top of the cinder cone is a view of nearby Punto Gallito and the palm oasis of Mulege to the north. The alignment of three nearby hills, Punto Gallito and two unnamed hills to the south, corresponds to the trace of a known mapped fault. Farther to the south, the length of Bahia de Concepcion and the entire Concepcion Peninsula including the Gulf Fault Zone are visible. For the next several kilometers there are many excellent, scenic views of Bahia de Concepcion and Concepcion Peninsula. Patches of cinders are exposed in the hills for the next several kilometers. These cinders are often utilized locally for road material.

123 Approaching Bahia de Concepcion the sparce flora consists of Palo Verde, Cardon, Garambullo, Mesquite, Busera, and Creosote Bush.

116.2 As the highway crests over a low pass, the parking area to the left provides an excellent view of Bahia Santispac, one of the most beautiful little bays along this coast. The light, milky green patch in the middle of the bay is due to shallow waters that barely cover a sandy shoal. The white sands in this area are not lithic sand but materials derived from the abundant calcareous fragments of marine invertebrates.

WHERE ARE THE SANDY BEACHES? Granitic rocks are coarse-grained and, as they weather, yield the sand-sized grains that compose sandy beaches. Because there are very few granitic rocks or granitic-derived sediments in northern Baja California Sur, there are few lithic sand beaches along the Gulf between Santa Rosalia and La Paz.

The volcanic sediments derived from the Comondu Formation which are prevalent in the southern part of the peninsula are fine-grained. When they weather, they yield smaller mud or silt-sized grains. In areas of volcanic sediments there are no real sand beaches. Hawaii is a good example of this phenomena. Most of the "sandy" beaches in this part of Baja are composed of bits and pieces of weathered shells and coralline material instead of granitic sand. This type of beach is called a biological beach.

BIOLOGICAL BEACH A beach is any sloping shore washed by waves or tides, especially the parts usually covered by sand or pebbles. Along Baja Sur's Gulf shoreline these sloping shores are not commonly covered by the sand which normally covers shores along the Pacific coast. The quartz and feldspar which form the major portion of most sands are too fine grained, in the volcanic rocks, to form sand. The majority of these "sandy" beaches are composed of mollusc shells, algae, and coral skeletons. These "sandy" beaches originated from biologically derived calcium carbonate and calcium phosphate particles that have been weathered to the same detrital particle size as quartz sand particles. Since the "sand" of these beaches were derived from the shells of once living molluscs and corals, they are said to be "biological sands". However, there are a few beaches in central Baja that are formed of quartz sand particles.

114.6 Just before the turnoff to Bahia Santispac, a fault is exposed in the roadcut. This fault roughly parallels the axis of the bay and may have been involved in the origin of the bay.

114.3 This side road leads to Bahia Santispac. This is one of Baja's most scenic beaches with its white sands and azure waters. The small red mangrove swamp, on the west side of the beach, is the northernmost one which is directly visible along the Baja Highway. The inlet around the swamp used to have numerous clams before it were "clammed out". Bird watching can be rewarding; pelicans, comorants, and frigate birds are among those birds commonly seen here.

BAHIA SANTISPAC AND MANGROVE SWAMP

The vegetation on the slopes to the inland side of the highway, appears to be sparse and dead because it matches the color of the metavolcanic soils. However, a closer look will reveal that they are growing quite well. It consists primarily of Cardon, Palo Verde, and Elephant Tree.

The Highway will pass numerous red mangrove swamps developed in the tidal creeks and estuaries which dot the Gulf coast from Mulege to the tip of the peninsula. The predominant plant of the swamp is the Red Mangrove (*Rhizophora mangle*), an evergreen shrub which form dense thickets in or near salt water or on alkaline soils. Soil and debris carried by the tides and stream runoff from the adjacent coastal environments, are washed among the network of stilt-like prop roots of the red mangrove and become caught.

Over many years this process slowly extends the shore line of the peninsula to the east into the Gulf. When fertilized, the white or cream colored flowers of the red mangrove produce a brown fruit which matures and germinates into a plantlet before falling off the parent. The plantlet has a long tap root, and when it drops off the parent plant, it falls like an arrow into the mud and plants itself. Several other plants commonly found in or near these swamps are Black Mangrove, White Mangrove, Pickleweed, Sand Verbena, and several species of native palms.

112 Posada (hotel) Concepcion on Bahia Tordillo. It has a general store, air and diving gear for rent, and a small airstrip.

109.2 The beach below and the beach of El Coyote are both biological beaches. The highway continues to pass a number of these beaches which are signed as "Playas Publicas" (public beaches), where anyone can camp. A fee per car per night may be collected for maintenance.

107 The seacliffs on the cove exhibit a weathering phenomenon called a wave cut notch (See 8:Pichilingue:14). A look into the water will reveal crystal clear waters with many changing hues on a bright day.

MAIN ROAD IN GULF

DRIVE IN THE GULF! Just south of El Coyote you can drive down to a small beach that was once located along the old road. When earlier travelers first drove down this part of the peninsula they found that the old road took the route of least resistance which skirted the shoreline in most places and climbing into the hills only when absolutely necessary. Below the new highway the old road was forced to skirt a steep sea cliff. There were a number of large boulders on the outside edge of the road about 4 meters from the base of the cliff. At low tide the highway, which passed between the large boulders and the cliff, was awash, and vehicles had to drive through 15 to 25 cm of water. At high tide the entire road and the large boulders were often covered by seawater. If you wanted to pass, you waited until the next low tide.

106.5 The highway turns inland at this point to bypass a difficult section of the coastline.

COVE ON CONCEPTION BAY

105 This beautiful fault line valley provides nice exposures of dark brown basalt and andesite breccias. One can really begin to appreciate the environmental extremes of Baja. The moisture dark-brown soils of the northern slopes along the right side of this valley support more vegetation, predominantly green barked Palo Verde, Cardon, and Elephant Tree, while the drier reddish soils of southern hillsides (left side) of this region are more sparsely covered with the desert species of Mesquite and Acacia. A few kilometers down the highway a cardonal of tall columnar Cardon cacti dominates the landscape.

102 The highway passes through a dense Cardon forest for the next several kilometers. Cardon associates are Elephant Tree, Palo Verde, and several other species of cactus.

100 The road cuts for the next few kilometers show the varied nature of the volcaniclastic rocks of the Comondu Formation.

99 Basalts form numerous barren talus slopes on the hills ahead. Talus slopes are a type of landslide which form when rocks fall, a few at a

time, and form a slope of debris which accumulates over a long period of time.

93.2 View of the tombolo of El Requeson, Isla El Requeson, and a red mangrove swamp. Herons and egrets are often seen along the edge of this swamp. Good clamming on the tombolo (sand bar) and snorkeling from the island have been reported. Its worth a visit.

TOMBOLO OF LA REQUESON

A **TOMBOLO** is a sand bar that has been deposited by currents and/or wave action between a mainland and an island or between two islands connecting the two landforms. At El Requeson a sand bar links the peninsula with Isla El Requeson.

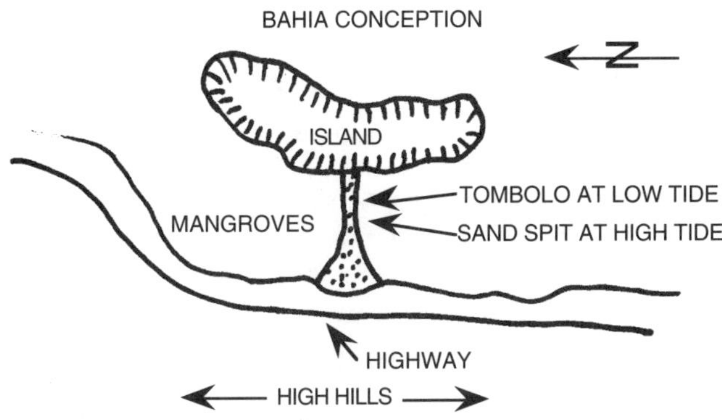

90 The view ahead is of Playa Armenta, the last of the biological sand beaches which have dotted the shoreline since Mulege.

89.7 To the left a stretch of the old road is visible immediately below the new highway. This is another reminder of the challenges that traveling the old road once presented. In the "old days" (pre 1972) as Baja travelers made their way along this rough road, one might smell pickles or mustard. It wasn't their imagination; one of the jar lids had come loose. The knowledgeable travelers immediately stopped to check before all was lost or ruined. Due to the constant jostling back and forth, labels would wear off of the cans and it was *"Pot luck anyone?"*

89 View to the left of Bahia de Concepcion. There are numerous, large alluvial fans on the far side of the bay on the Concepcion Peninsula. These large alluvial fans were formed by streams which carved material from the hills of the Concepcion Peninsula. There are no large fans on this side of the bay. This combination of numerous large fans on one side and a few small fans on the other is good evidence for the recency of movement on the Gulf Fault Zone which parallels the western side of the bay. The movement resulted in a down-drop of the western side while the eastern shore was uplifted. The large fans on the east side had a much longer time to form.

83 The cinders in this roadcut are part of a Miocene cinder cone which is one of several exposed along the highway over the preceding five kilometers. These cinder cones are probably distributed along a branch of the Gulf Fault Zone.

CARDONAL AND PLIOCENE BEDS

76.6 Numerous very large Cardons can be seen growing in a dense stand known as a cardonal on the well-drained inland slopes. A quick look to the left shows that they rapidly become smaller and sparser toward the shoreline of the bay; this is due to the more saline soils and poor drainage. Cardons grow best on well-drained slopes. Dense stands of *Bursera*, Copal, and Jumping Cholla are associated with the Cardon.

73.5 The flat-lying, whitish to pinkish rocks to the left in the far distance in a gap in the hills are marine Pliocene strata consisting of limestones, tuffs, and redbeds. They were formed in association with hot spring deposits.

73 As the highway ascends a grade, the view to the rear is the last one of Bahia de Concepcion and the Concepcion Peninsula. Its beauty will be remembered on the long drive through the desert as the highway proceeds to the south and the Cape Region.

71 The highway descends through roadcuts which expose the andesite mega-breccias of the Comondu Group.

68.1 A large dark basalt dike cuts the light pinkish mega-breccia.

The vista opens to the south onto the Loreto Graben. The alluviated valley which the road follows is a graben bounded on the right by the high eastern escarpment of the Sierra de Giganta and a horst composed of volcanic rocks to the left. The highway will continue to follow the axis of this graben all the way to Loreto.

SIERRA GIGANTIA FROM PASS

62.4 The highway passes through the palm oasis of Rosarito, the third Rosarito along the highway.

59.9 The graded road to the right leads to the towns of La Purisima, and Comondu.

56.8 The low hill 0.5 kilometers to the right is cut by two prominent nearly vertical dikes which are offset in a right lateral sense by one of the numerous traces of the Gulf Fault Zone.

55.1 The rancho on the left is Ascencion. Another road to the right leads to the Pacific side of the peninsula via Canipole (8 kms.), Comondu (76 kms.), and La Purisima (74 kms.). This is as far south as most people drove before the road was paved because the old road continuing on to Loreto was a dead end and used only by people going there.

54 The prominent white hill approximately one kilometer west of the highway is a hypabyssal intrusive plug along the Gulf Fault Zone.

53.5 El Bombedor, another small palm oasis, is a small road side Rancho typical of this region.

36 There are ranchos on the alluvial plain. Aerial reconnaissance and photographs show that most of these ranchos are located close to linear fault features clearly marked by lines of water-loving, more mesic vegetation. The faults in this region create a zone of "crushed" rock which allows ground water to come near the surface.

Between Kilometers 36 and 32, the hills to the left contain numerous north-south fault traces. Several fault scarps and small horsts and grabens are readily apparent. These faults are part of the Gulf Fault Zone which passes through the Loreto embayment and Bahia de la Conception.

35 Directly west, the high peak of the range is Cerro la Giganta elevation 1796 meters.

A MYSTICAL LAND CALLED CALAFIA: As the Spanish explored Mexico they brought with them a book with legends of a mystical land called Calafia, an island somewhere to the west of Spain. Throughout their explorations the Spanish looked for the land of Calafia which supposedly was occupied by large "Amazon" women, known as Gigantis. They were said to have lots of gold and jewels, and thus, the real motivation of the search of the Spanish! When the Spanish traveled from Mainland Mexico across the Gulf of California to Baja California, they thought they had found the island of Calafia. It was not until much later that it was proven to be a peninsula. It is quite possible that California was named for the mystical island of Calafia, not for the Mexican word, Californax (which means "hot furnace"). When the Spanish first settled in this area, they imagined they could see one of them reclining in the Sierra de la Giganta.

On the fan below the peak is a fault scarp which offsets the fan. The fault parallels the highway halfway down the fan with approximately 10 m. of offset and is marked by a low line of hills which run through the fan (Look at the line at the top of the tallest Cardon in the photo). Further to the left is a series of small, sharp-crested hills. These are granodiorites, dated at 145 and 87 m.y.

CERRO LA GIGANTIA AND A FAULT SCARP

30.5 This turnoff leads to the small date palm community of San Juan Bautista. To the southeast is the Boca San Bruno. Early Jesuits attempted to establish two visiting stations for the Loreto Mission in this area: Mision Guadalupe de San Bruno, 1683, and Mision San Juan Bautista Londo, 1687. Both attempts were unsuccessful.

29.5 The highway climbs through the first of the marine Miocene-Pliocene sandstones, siltstones, and conglomerates of the Loreto Embayment.

A MARINE TECTONIC BASIN: Major movement of the Gulf Fault Zone appears to have initiated the formation of the Loreto embayment of Miocene(?)-Pliocene age; it is a thick sequence of conglomerate, sandstone, mudstone and limestone with pecten and oyster reefs, tuff, andesite and basalt. Tuff beds and a lahar indicate volcanic action during the filling of this basin. At least one major angular unconformity and the lensing of conglomerate beds suggests intermittent deformation. Basalt (6.7 million years) intrudes and overlies part of the section which may be older than the rest of the basin. Other dates of 1.9, 2.1 and 3.3 m.y. have been obtained from tuffs interlayered in the marine sequence (Mclean, 1987, 1989). The syncline of the Loreto embayment is folded into a N-S anticlinal structure with numerous N-S and NW-SE trending faults which may have one to two thousand meters of cumulative displacement. The conglomerates contain granitic and metavolcanic clasts from outcrops to the west.

The Loreto embayment is one of the few tectonic basins in the southern part of the peninsula. The darker brown flat-lying Miocene rocks which cap the mesas to the west are as young as 10 million years old. This embayment was formed after faulting warped and uplifted the mesas. One dike, dated at 6.7 million years, cuts the sedimentary rocks. At this time the northern part of the Gulf of California had already formed, and the more southerly parts of the gulf were being formed had moved to the south. The Loreto embayment was warped downward and received sediments from the surrounding highlands. The conglomerates indicate intermittent uplift of the same areas and more powerful streams. The yellow beds indicate wearing down of the source areas with less powerful streams and only finer material is carried far into the basin. The area was continuously faulted. Several tuff beds are exposed in the section which indicate continuing volcanism. These rocks are highly fossiliferous.

28.9 Evaluation of the stratigraphy seems to indicate that the strata on this roadcut exposes andesite dikes which intrude into Pliocene beds. The prominent dike in the middle yielded a 6.7 m.y. K/Ar (potassium/argon) date. However, tuffs in the arroyo nearby yielded a 1.9 m.y. K/Ar date.

27.3 As the highway crosses a large arroyo, an autobrecciated, andesite lahar is exposed on both sides of the highway.

26.5 Tuff is being quarried for use as building material from a white tuffaceous sandstone outcrop on the hill above the highway.

26 An oyster shell reef continues for nearly a kilometer to the south along the highway.

25.3 The oyster-shell reef and white tuffaceous sandstone parallel the highway for a short distance before crossing under the highway to the right to interrupt the yellow sedimentary beds exposed there. The surface of the hill on the right is veneered by oysters of the genus *Ostrea*. The tuff yielded a 2.1 m.y. K/Ar date.

22.9 A small rancho is located along the highway. The yellow beds are well exposed in the hills near this rancho. They represent a time of low sedimentation. Benthonic foraminifera from these beds indicate that they were deposited at shelf-slope break depths. The conglomerates exposed here represent periods of uplift of the adjacent highlands along the faults in the Sierra la Giganta.

20.1 This dirt road leads to San Juan Londo. The vista to the rear provides a good view of the Gulf. Just south of the water is a small, rounded volcanic hill. This was the site of a small visiting station of the Loreto mission. There are approximately 300 meters of conglomerates and sandstones between the yellow beds in this area.

22 The road up the sandy arroyo, to the west, leads to outcrops of granitic rocks.

PLIOCENE MARINE SEDIMENTS

19 To the southeast is the southern part of the Loreto embayment. The volcanic hills to the left side of the highway are overlain by the white tuffaceous sandstones and gray conglomerates of the lower part of the Carmen-Marquer Formation undifferentiated. These beds are thousands of meters thick and contain abundant fossils. Common fossils include: whale bones, *Argopecten sp.*, *Ostrea titan*, *Turritella sp.*, *Chama sp.*, *Chione sp.*, and numerous other species of Mollusca (clams and snails). The southern part of the Loreto

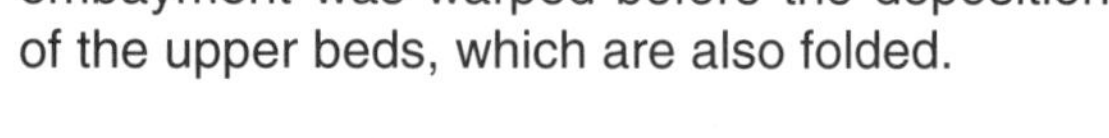

embayment was warped before the deposition of the upper beds, which are also folded.

18.5 A pecten reef is to the right above the highway.

18.3 There is a view to the left into a distant amphitheater of yellow beds overlain by conglomerates with an isolated thumb of conglomerate in the center. The yellow beds have yielded very large oysters and whale remains.

13.8 The highway descends into a small arroyo. The limestone on the ridge to the left is much younger than and unconformably rests on the tilted edges of the Loreto sedimentary rocks. There is a pecten reef in the roadcut opposite the guard rail.

12.8 The highway crests a small grade and begins to follow Arroyo de Arce.

The white volcanic tuff bed seen in the roadcut is repeated five times for a cumulative separation of 200 meters on this bed alone. Sand dollars and pectens have been collected from the layer above the tuff bed.

FOSSIL SAND DOLLAR

12 Blocks of white tuff were mined at Rancho Las Piedras Rodadas to the right. This tuff is 3.3 million years old.

10.7 The highway crosses the crest of an anticline. The conglomerates and Sandstones now dip to the south.

9.6 Note the prominent fault which dips 45 degrees to the north.

9.3 This side road leads to Microondas Loreto on volcanic rocks which yielded a 14.9 m.y. K/Ar (potassium/argon) date.

9 These roadcuts reveal numerous small faults and scattered pecten fossils. The highway crosses the deep Arroyo de Gua. This arroyo passes through a thick faulted section of the Loreto sedimentary rocks.

6.5 First good view of Loreto and Isla del Carmen.

6 The vegetation consists of Cardon, Pitaya Agria, Pitaya Dulce, Palo Blanco, Palo Verde, Acacia, Elephant Tree, and Leatherplant. One of the commonly seen birds in this area is the large spotted-breasted Cactus Wren (*Campylorhynchus brunneicapillus*).

The **Cactus Wren or Matraca Grande** is the largest wren of the deserts and arid hillsides throughout the entire Baja peninsula. They are commonly seen feeding and nesting below 1200 meters among thickets of thorny scrub, large cacti, clumps of yucca, or mesquite. Most wrens are small, restless brownish birds with finely-barred narrow rounded tails that are held erect. The Cactus Wren is easily identified in the field by its streaked back, wide, white eye stripe, and heavily spotted breast. Their bulky nests are easily identified, tucked into cholla cactus or thorny bushes. When it feeds, the Cactus Wren furtively searches through the ground litter for small invertebrates, usually insects.

3 The 30 kilometer long island which is visible immediately offshore is Isla del Carmen.

ISLA DEL CARMEN is surrounded by water with depths of less than 200 meters and exhibits greater water clarity than found along the peninsular shores. Increased clarity is due to lower plankton densities and smaller quantities of suspended sediments. Geologically, Isla del Carmen is composed of Miocene volcanics of the Comondu Formation and marine Pliocene rocks of the Salada Formation.

There are 18 species of plants endemic to the Gulf islands. The Barrel Cactus *Ferocactus diguetii* var. *carmenensis* is the only endemic of Isla del Carmen. Botanically, the predominant plants of Isla del Carmen are Agave, Cholla, Elephant Tree, Hedgehog and Nipple cacti.

Isla del Carmen is inhabited by employees of a solar salt works which has operated since the Spanish first settled Baja California.

0 The tower of Mision Nuestra Senora de Loreto can be seen; it is surrounded by palm trees in the main part of Loreto.

To the north, is the crest of the anticline which forms the east side of the Loreto embayment. The dark-colored hills on the crest are volcanic rocks, circled by a lighter limestone with the sandstones and conglomerates of the Loreto Formation in the lower hills. Isla Coronado is visible on the skyline.

GULF DESERT FLORA ON ISLA CORONADO

ISLA CORONADO is a very recent volcano with some flows still devoid of vegetation. This island contains a Pliocene marine terrace and a large cobble beach bar at 10+ meters elevation. Many flows have moved down canyons and are truncated by the ocean at their base. There are no terraces on the sides of the island with the fresh flows. The marine terrace indicates that the island is at least 125,000 years old. The truncation of the more recent flows by the sea suggests that the most recent volcanism was at least a few thousand years before present.

MISION NUESTRA SENORA DE LORETO

LORETO is Baja's first city. Exploration of the Baja peninsula began in 1535 when Hernan Cortez established a small colony on the present Bahia de La Paz which was abandoned in 1537 due to supply shortages. Exploration of Baja was continued by Hernando de Alarcon (1540), Juan Rodriguez Cabrillo (1542), Sebastian Vizcaino (1596), Nicholas de Cardona and Juan de Iturbe (1615), Francisco de Ortega (1664 and 1666), and Francisco de Lucenilla (1688). In 1697 Padre Juan Maria Salvatierra with six soldiers landed on Baja's east coast and established Mision Nuestra Senora De Loreto at the delta of a large perennial stream, Arroyo de Las Parras. For the next 132 years Loreto was the capital of Baja California. In 1829 a hurricane destroyed Loreto and the capital was moved to La Paz. Flood waters raging down Arroyo Las Parras have destroyed the town on eleven occasions since 1697. Recently, some efforts have been made to control the flooding with channels. The town has also been damaged by earthquakes. A particularly large one occurred in 1877.

The mission in Loreto is the first and the oldest in Baja California. It was from here that Father Junipero Serra began in 1769 to establish the missions in Alta California. The church was completed in 1752, destroyed and rebuilt. The Loreto mission museum is worth a visit. Today Loreto is the seat of the municipal government and a sport fishing center. Good hotels are located in Loreto and all services are available.

CATTLE ON SAN JAVIER ROAD

CUEVA LAS FLECHAS
SIERRA SAN FRANCISCO

Log 7 - Loreto to Constitucion [145 kms = 90 miles]

The Highway continues south on the narrow Gulf coastal plain between the rugged Miocene volcanic mesas and the Gulf of California with its many faulted volcanic islands. South of Ligui the Highway climbs a grade through the Miocene volcanics and the Gulf fault zone to the mesa tops. It then follows a gentle canyon in the mesas to the flat Magdalena Plain with metamorphic islands offshore.

120 LORETO TURNOFF: The uplifted fault scarp cliffs of flat-layered volcanic tuffs and sedimentary rocks to the west are the main massif of the Sierra la Giganta. From this point it is hard to realize that there is a relatively flat, gently sloping, tableland on top of the massif which extends west to the Pacific Ocean. The prominent peak to the right of the large pass in the Sierras is Cerro Pelon de Las Parras, a shallow intrusive andesite plug which yielded a 19.4 m.y. K/Ar date.

The low hills to the northwest are composed of 94 m.y. old metavolcanic rock and 143 to 87 m.y. old tonalite with a moderate to complex structure intruded by a dike system. The presence of older granitic and metavolcanic rocks at this point in the Gulf of California plus the lack of Peninsular Range basement in south-central Baja confirms geophysical evidence for the continuation of the Cretaceous syncline under a major part of the State of Baja California, Sur.

South of Loreto the Gulf Fault Zone and the highway converge until the highway is forced to follow a narrow hilly coastal plain between the Gulf and the steep cliffs of the Sierra la Giganta.

117.6 The road to the right leads 48 kilometers up Arroyo Las Parras into the ruggedly scenic Sierra la Giganta to Mision San Francisco Javier de Vigge. This mission, which was founded in 1699, was the second Jesuit mission established on the peninsula. Some who have visited Mision Javier consider it the finest example of mission architecture in Baja. It has the distinction of being the only original mission church that has remained intact. All of the others have been destroyed and rebuilt at least once.

The road crosses a Comondu section which is different from the other sections crossed by the main highway at San Ignacio and south of Loreto.

MISION SAN FRANCISCO JAVIER DE VIGGE

116.6 Turnoff to the Loreto International Airport.

The highway continues south along the narrow coastal plain between the abrupt escarpment of the Sierra la Giganta and the warm waters of the Gulf. The escarpment was a formidable barrier to the early missionaries. They finally breached it at San Javier. Isla del Carmen dominates the Gulf and the northern portion of uninhabited Isla Danzante is just visible to the southeast.

In this region the highway was constructed in the Gulf Fault Zone. The disjointed and irregular hills in the foreground are in the fault zone. The crushed rocks adjacent to the highway do not exhibit bedding, while impressive flat bedding is well exposed in the high eastern escarpment beyond the fault zone to the west.

111.1 This turnoff leads to Nopolo Beach and a resort with a golf course. A beautiful small cove and mangrove swamp are located near a tombolo that connects the peninsula to a small rocky island. A tombolo is a sand spit which connects an island to a mainland. At some high tides this tombolo may be awash. The scenic beach in this cove is a rarity in this area. Lithic sand, instead of crushed shells, forms this beach.

At Nopolo winter visitors may be treated to a spectacle of wildly-diving Brown Pelicans (*Pelicanus occidentalis*)! During the winter currents and winds combine to pile up schools of small fish along the coast of this area. The pelicans take advantage of the conditions and become embroiled in feeding frenzies.

BIRD NAME	LIKELY LOCATION
Amer. White Pelican	Gliding along the shore
American Kestrel	Wires and fence posts
Cactus Wren	On cacti
Calif. Brown Pelican	Gliding along the shore
California Quail	On the ground
Costa's Hummingbird	Feeding on red flowers
Gila Woodpecker	In the scrub
Gray Thrasher	On the ground looking for food in the litter
Greater Roadrunner	Crossing the highway
Ladder-back Woodpecker	Flitting in the air
Pyrrhuloxia	Mesquite and oak woodlands
Red-Tailed Hawk	Tops of poles and fence posts
Turkey Vultures	Soaring in the sky
Western Meadowlark	Fence posts and fence wires
Xantu's Hummingbird	Feeding on tubular flowers

The **California Brown Pelican or Pelícano Moreno** is one of the large aquatic fish-eating birds seen along both of Baja's coasts and on many of the Gulf islands in Baja where they nest in large colonies. They are commonly seen flying in long straight lines only centimeters above the surface of the water. Feeding is accomplished by diving into the sea from as high as 50 m. The slow deliberate flight of the brown pelican, which is low over the water with sudden plunges for fish, makes its identity unmistakable. The Brown Pelican is commonly seen on piers and docks.

PELICANS AND GULLS

107.5 Loreto and its white church tower are seen in the distance to the rear.

Many species of birds inhabit the Gulf Coast Desert. Some of the more common birds of this area are listed below:

The highway ascends a small grade through dark basalts with a lagoon, inlet, and the small mangrove swamp of Nopolo on the left. A number of formerly inhabited caves have been found in the Sierra la Giganta to the east of Nopolo. They were occupied by Indians who subsisted on the clams and other marine life of the cove.

VIEW FROM INDIAN OCCUPIED CAVE AT NOPOLO

103 To the left is Playa Notri, another lithic sand beach where many species of game fish, which include Marlin, Sailfish, Sierra, Dolphin Fish, Yellowtail, and Grouper, often come close to shore.

99.5 **STOP here to enjoy the view!** A roadside rest area is on the left side of the highway where there is an excellent view of the little cove of Juncalito. Juncalito is Spanish for little tilted rock.

The water in the low pass to the southeast is part of Bahia Escondido which is located between the peninsular coast and Isla del Carmen. The peaks near Punta Candaleros to the south of Puerto Escondido are andesite breccias and lahars.

To the south in the distance is Isla Danzante composed of faulted Miocene andesite breccias and volcaniclastic sedimentary rocks. The north end of Isla Monseratte, composed of Miocene volcanic rocks and Pliocene marine sedimentary rocks, and Isla Santa Catalina to the left, composed of Mesozoic granitic rocks, can be seen in the distance. The largest island, Isla del Carmen, composed of the Miocene volcaniclastic and Pliocene marine rocks, is directly offshore to the east. To the north is the Pliocene to Recent basalt cone of Isla Coronado (*See* Anderson, 1950, for a description of the geology of the islands).

99.3 There is a large solitary Wild Fig Tree (*Ficus palmeri)* which grows on the beach at the end of the gravel bar to the left.

WILD FIG ON BEACH

WILD FIGS are members of the Mulberry family. They range from the palm oasis of San Ignacio south to the Cape region where they are usually seen growing alone on rocky cliffs, in canyons, and occasionally on gravel beaches such as this one. Wild Figs produce a barely edible dry fig in the late spring and early summer.

97.5 This is the entrance to Juncalito Cove (Refer to Km 99.5). This is an great place to stop and watch for birds. One that is especially nice is the Crimson Pyrrhuloxia.

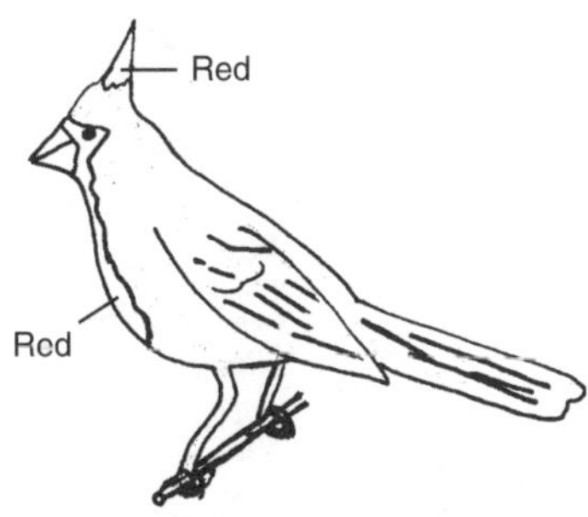

The Crimson **Pyrrhuloxia or Cardonal Torito** inhabits the southern half of Baja where they are known as the gray bird of the cardon cactus. They are a beautiful crimson and gray colored cardinal relative with a yellow, straighter parrot-like bill. Pyrrhuloxias are commonly seen in the mesquite scrub along the road to the gulf's Bahia San Luis Gonzaga and surrounding Loreto where they usually feed on the ground, taking seeds and insects. This bird derives its name from the Greek, meaning crooked-billed red finch.

94.3 **PUERTO ESCONDIDO:** This paved side road leads 2.4 kilometers to the port of Puerto Escondido. Steinbeck in his "Log of the Sea of Cortez" claimed this as his favorite place in the Gulf. The author of this guide has spent many weeks here and agrees. This area is worth a visit. It exhibits many excellent tide pools and the coastal waters abound with fish and other sea life. Twenty-five pound Yellowtail have been caught by the author off the pier.

A fault is exposed in the saddle directly in front of the pier with dark basalts on the near side and fluviatile andesite breccias on the far side.

Magnificent Frigate birds are commonly seen soaring nearby.

FRIGATE BIRD

The Magnificent Frigate Birds or Fregata Magnífica are probably the most aerial of all sea birds and thus present a common but thrilling sight over the waters along the gulf coast of Baja. These birds possess the greatest wingspan (up to 2.5 meters), in proportion to their body, of any known species and are hardly equaled by any bird in their powers of sustained, soaring flight. Their most identifiable field characteristics are the prominent crook in the slender, streamlined wing, the long slender forked tail, the way they soar high in the sky, apparently hanging motionless, their steep, swift dives to snatch fish from the sea, and the naked red inflatable throat (gular sac) of breeding males. Immature birds have white on their heads and throats. They often act in an aggressive and piratical manner, chasing and colliding with blue and brown-footed boobies, terns, and gulls, upsetting them so much that they regurgitate their last, fishy meal, which the Frigate then aerobatically snatches from midair. No doubt the name Frigate (a pirate ship) was derived from this behavior. Another method of obtaining food is to steal chicks from tern colonies or as a last resort to fish for themselves. Frigates nest on offshore islands in large untidy stick structures placed in anything from bush to a mangrove tree to avoid predation by the four footed predators that inhabit the nearby peninsula.

At night during the right season, you may make eerie sightings of sea lions which swim in a sea which abounds with a microscopic phosphorescent bioluminescent protozoan of the genus *Noctaluca.* The bluish-green light of *Noctaluca* is strong enough to read by.

93 The view offshore is dominated by Isla Danzante. Isla Carmen is to the left and Monseratte is to the right. The three small isolated "rocks" in the bay to the right are Los Candaleros (the Candles). Los Candaleros are isolated pinnacles known as stacks (*See* 9:6).

89 The view to the rear is of the uplifted steep eastern escarpment of the Sierra la Giganta. The flat and well-bedded strata extend from the Pacific eastward across the peninsula to the edge of the scarp. In the fault zone the beds become chaotic, crushed, and deformed.

86 The highway continues south along cliffs of fluvial volcanic breccia which exhibit large clasts and numerous caves. Isla Danzante consists of this same material cut by dikes and faults.

SCARP ALONG GULF FAULT ZONE

84.2 This turnoff leads to Ligui and a beautiful cove with a rare lithic sand bottom which can be reached by this road.

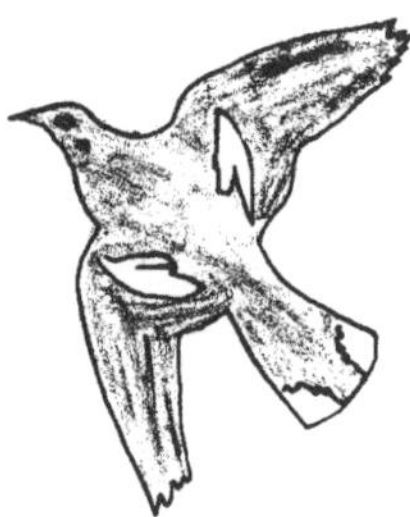

South to the Cape Region, White-winged Doves (*Zenaida asiatica*) are commonly seen on the ground or flying across the highway.

White-Winged Doves or Paloma de Alas Blancas are the only doves with large white wing patches which contrast with general olive-brown colors of the body. Although they are widespread throughout the peninsula, they are most commonly seen in thorn scrub, mesquite groves, Riparian Woodlands, and the open dry desert areas south of Loreto. They have a drawn out, cooing call, "who-cooks-for-you," in many variations.

The flora of this area is dominated by *Acacia*, Garambullo, Pitaya Agria, Pitaya Dulce, Cardon, Mistletoe, Palo Blanco, Palo Adan, and the Vining Bougainvilla-like San Miguel. The disturbed soils around the rancho support Jumping Cholla

83.5 The highway turns through a wash, reaches the bottom of the grade, and begins the ascent of a steep grade south of Ligui. A fault shear which is part of the Gulf fault zone is visible in this roadcut.

78.8 There is a quick view to the left of the Gulf and the small village of Ligui. Islas Monseratte and Carmen lie directly offshore. The low volcanic cone of Isla Coronado is just visible to the north.

Between here and the top of the steep grade, the highway passes over the major Gulf Fault Zone. Most of the major motion along this fault was pre-Miocene. There is evidence of lesser Miocene to Recent motion. The zone is difficult to identify because it is wide and diffuse. The complexly faulted areas, offset, and deformed bedding of this region is generally marked by a foothill belt rather than the undeformed well-defined bedding of the mesas of the Pacific slope.

75.7 This turnout provides a good view to the north of the majestic massif of the Sierra la Giganta. The high crest of the range with the undeformed sedimentary rocks which extend from the western Pacific slopes drops rapidly to the east into the undulating fault-deformed foothills adjacent to the highway on the right.

73.7 The large orange-red splotches on the low hill to the left are crustose lichens.

70.8 As the highway crests the top of the Ligui grade, there is a road to the east which leads to Microondas Ligui. The highway descends through an east-west trending canyon down the gentle western slope of the Sierra la Giganta for tens of kilometers.

GULF FAULT ZONE AT TOP OF GRADE

69 The undeformed well-bedded strata of the Pacific drainage of the Sierras are now very obvlous along the canyon walls.

64 This side road leads 41 Km to a small fishing village on the gulf at Bahia Agua Verde and Punta San Marcial.

The cliffs on the left are covered with two obvious kinds of lichen (Fungus with an intracellular mutualistic single celled algae of the genus *Chorella*). The most abundant kind is green, the less obvious ones are orange-red (rust) colored.

60 The Cardon cardonals are extensive and are primarily located on the cooler moister north-facing slopes, while Mesquite grows abundantly in the well-drained canyon bottom. The two sides of the canyon show notable differences in vegetation typical of north and south-facing slopes.

56 The vegetation of this region represents an ecotonal zone (*See* 2:118) of a mixture of plants of the Magdalena Plain desert area and the Gulf Coast desert area. It is dominated by a Cardon-Palo Verde forest with scattered Pitaya Dulce, Acacia, Cholla, Beaver-tail, vining San Miguel, Palo Blanco, Palo Adan, Purplebush, Lomboy, and Elephant Tree.

54.5 On the hill to the south, brown basalt caps the lighter gray-brown andesite breccias, lahars, and fluvial sedimentary rocks of the Comondu Formation. They dip gently to the west. The highway descends at approximately the same gradient as the arroyo itself. The gradient of the arroyo is controlled largely by the slight dip of the Comondu Formation toward the Pacific Ocean. The arroyo descends for some distance along a single bed, drops down through a few beds, and then flows along another bed. Marine Miocene rocks are not exposed along the highway as it crosses the peninsula in this region.

54 Andesite breccias, typical of the Comondu in this area, are exposed in this roadcut and on the opposite wall of the arroyo.

49 More andesite breccias are exposed in the opposite wall of the canyon. These monolithologic andesite breccias take the form of lahars with andesite flows, flow conglomerates, and fluvatile conglomerates. They grade into finer sandstones and well-bedded sedimentary rocks. This is a facies of the Comondu Formation.

44 These outcrops are fluvatile sandstones and conglomerates of the Comondu Formation capped by basalt.

The highway leaves the canyon and passes onto the Magdalena Plain. It is largely composed of Miocene and Pliocene rocks overlain by the same type of Pliocene-Pleistocene limestone which caps the Vizcaino Plain to the northwest. These sedimentary rocks are only occasionally exposed in the arroyos or road metal quarries of the Magdalena Plain. Bahia Magdalena is a deeper remnant of this shallow embayment.

On a clear day the Pacific islands of Isla Margarita and Isla Magdalena, which are composed of Franciscan metamorphic rocks, are visible ahead to the left. These islands are the remnants of the sea floor of the Pacific Plate which was subducted under the North American Plate. Radiometric ages obtained from these rocks seem to cluster around 135 m.y. They are very similar to the Coast Range Province of California.

43 The Magdalena Plain Desert flora is represented by large cardonals with associated Pitaya Dulce, Palo Verde, Acacia, Wax Plant, Pitaya Agria, Garambullo, Candelabra Cactus, Greasewood, Mistletoe, Jumping Cholla, and Leather plant. Epiphytic Ball Moss (*Ramalina*) is occasionally seen growing on the tops of the Palo Verde.

28 Zona de Neblina. Zone of the fog. The fog moisture in this zone enable an abundance of epiphytic gray-green Foliose Lichens and dark Ball Moss to grow on the shrubs and cacti.

20 This area is much drier than that near Kilometer 43 and supports significantly fewer shorter scattered Cardons, Pitaya Dulce, Candelabra Cactus, Ball Moss, Creosote Bush, Lomboy, Palo Verde, and Palo Adan.

15.5 The size of the Que Reparrio bridge attests to the fact that on some occasions a large quantity of water flows over this plain.

0 At Ciudad Insurgentes the highway turns sharply to the southeast. The kilometer markings change to 236 at this junction and descend as the highway approaches La Paz.

225 The area is heavily vegetated by native Pitaya Agria, Lomboy, Mesquite, Creeping Devil Cactus, Cheese Bush, several species of Cholla, Lichen, Torote, Cardon, Palo Adan, and Old Man Cactus. The vegetation of this region is characteristic of the very dry Magdalena Plain region of the Desert Phytogeographic Province of Baja California. This extremely dry region of Baja receives only 2 to 5 Cms. of precipitation annually.

212 **Ciudad Constitucion** is a thriving agricultural community and is the gateway to Bahia Magdalena and Puerto San Carlos.

CUEVA PINTADA
SIERRA SAN FRANCISCO

Log 8 - Constitucion to La Paz [212 kms = 132 miles]

The Highway crosses the flat Magdalena Plain, with views of the offshore islands, to Santa Rita. From this point the highway begins to traverse a series of gentle washes and dissected mesas in marine sedimentary rocks to San Agustin where it climbs onto a dissected surface of fluvial volcanic sedimentary rocks and crosses the peninsula. Near La Paz the highway descends the Gulf Scarp through the fluvial volcanics to the alluviated La Paz Plain.

212 **CONSTITUCION** is a large agricultural community. This is one place where the biblical prophecy "and the desert shall bloom as a rose" has come to pass. Farming has been made possible by deep artesian wells drilled by the Mexican Government which bring "fossil water" to the surface from ancient aquifers. The amount of water withdrawn from these aquifers through wells is carefully monitored to prevent the over usage and eventual depletion of this limited resource. Some water is naturally available from infrequent summer rainstorms or late summer-early fall chabascos or hurricanes. These produce sudden downpours which may overflow the arroyos in a matter of minutes. However, much of this water is not available to the flora of the Magdalena since it runs off quickly before much of it has a chance to soak to the roots of the plants.

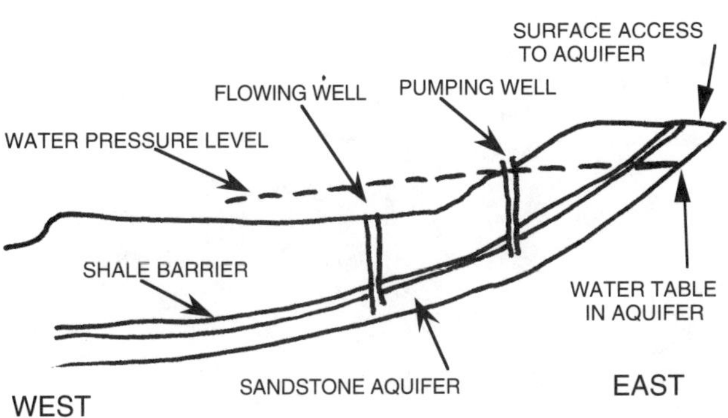

198 The village of Via Morelos is located in the middle of a farming area. Most of the nonagricultural regions of the Magdalena Plain support disturbed vegetation because the land has either been cleared or overgrazed.

190 Mision San Luis Gonzaga is located in the low foothills east of the Magdalena Plain. The mission was formerly the site of a garrison of troops used to reinforce the sailors on the Manilla galleon when it came through the Bahia Magdalena area. This mission is situated along a stream that follows a transpeninsular fault offsetting the Cretaceous geosyncline by a total of 50 kilometers in a right lateral sense. The fault has deformed lower Miocene sedimentary rocks and is overlain by flat-lying 20 million year old volcanic tuffs. A high clearance vehicle can reach the mission from this junction.

MISION SAN LUIS GONZAGA

173.8 The quarry on the right exposes limestones of the fossiliferous Miocene-Pliocene Salada Formation. This limestone can be seen over the Magdalena Plain. There is a graded road just south of the quarry which leads 32 km to Cancun on Bahia Magdalena.

This plain resembles the Great Valley of California. Structurally, the two regions are nearly identical. The highway roughly follows the axis of the Cretaceous geosyncline, as defined by geophysical surveys, and is underlain by up to 20,000 meters of Cretaceous and Tertiary sedimentary rocks.

170 The vegetational cover between this Km. and Santa Rita is sparse. Depending on the location the flora of this area is alternately predominated by Cardon, Cholla, Elephant Tree, Palo Verde, Palo Adan, Cheese Bush, and Creeping Devil Cactus, annual composites, Leatherplant, Greasewood, Garambullo, and Pitaya Dulce.

On clear days the islands of Isla Santa Margarita and Isla Magdalena can be seen to the west across Bahia Magdalena. These islands are composed of Franciscan-type rocks similar to the Coast Ranges of California. They contain thrust sheets, pillow basalts, serpentinite, and garnet hornblendites. These rocks have been K/Ar dated at 134 m.y. (Minch, 1971; Yeats, *et al.*, 1971).

ELEPHANT SEALS ON ISLA MAGDALENA

AMPHIBOLITES ON ISLA MARGARITA

157 From Santa Rita the highway traverses the gentle Pacific slope drainages through a series of washes and dissected mesas.

For the next several kilometers the highway passes through the reddish-brown and brownish-yellow beds of the Tepetate Formation.

The washes are populated by dense mesquite, while the hillslopes support Creosote and sparse Leatherplant and Palo Adan with stands of Jumping Cholla and Pitaya Dulce.

144.3 This roadcut exposes a limestone which has been mapped as part of the Miocene-Pliocene Salada Formation. The Salada Formation (Heim, 1922) is named for marine Pliocene and Pleistocene sedimentary rocks in Arroyo Salada. Judy Smith (1984) now feels that the type section of the Salada Formation ranges from late Miocene to early Pliocene in age and represents a shallow marine embayment.

143 The floral species which cover the Pacific slopes of the Sierra de la Giganta are typical desert dominants of the Magdalena Plain area of the Desert Phytogeographic Region. They continue to be represented by epiphytic Ball Moss and *Ramalina* lichen which grow on the elephant trees *Bursera microphylla* and *B. hindsiana*, Palo Adan, Leatherplant, Cardon, Jumping Cholla, Pitaya Dulce, and Garambullo.

135 The highway makes a long crossing of an arroyo. Shark teeth have been found in the Miocene exposures located below the El Rifle microwave tower to the left of the highway.

EL RIFLE

127.1 This is the turnoff to Rancho San Pedro de la Presa. It heads east to Mision La Presa, north to the short-lived Mision San Luis Gonzaga (1737-1768), and back to Ciudad Constitucion. The dirt road traverses a cross-section of Miocene exposures in this part of the peninsula.

126.2 The bridge at Puente Ventura was washed out in a flash flood in 1978. Future washouts may occur since the highway crosses a wash that periodically and briefly conducts enormous amounts of water from the infrequent sudden downpours which occur during rainstorms and chabascos that are common in this region.

THE LEGUMINOUS ACACIA: There are more than 900 species of Acacia trees and shrubs in the world with 64 species found in Mexico. Only about 20 species of Acacia occur in Baja from the northeastern Desert Region to the Cape Region. The fruit of the Acacia is a pod (a specialized folded leaf) with several seeds that are used by both man and beast. Indians ground the seeds into a meal for food. Cattle also relish the pods. In regions of North America where cattle are moved in "drives", Acacia is often found growing in straight lines due to cattle being "driven" in "straight lines" and "planting" the seeds as they defecated along the drive.

112 The prominent peak to the east is Cerro Colorado (450 meters).

CERRO COLORADO

Mesas capped by dark brown and gray-green volcanic rocks can be seen to the left. Immediately below, sandy yellow sedimentary rocks are visible in the middle of the section, and shaley yellow beds are at the base of the section. These rocks were originally referred to the Isidro, San Gregorio and Monterey Formations by earlier workers. They are now renamed and are regarded as members of the El Cien Formation of Applegate.

These Miocene formations are well exposed in various cuts in arroyos over large parts of the Magdalena Plain. The plain is so flat that few rocks are exposed outside of the arroyos.

108 For the next several kilometers the highway passes through the yellow beds that have been mapped as the Isidro Formation. These are near-shore facies, but not quite as near-shore, nor as tuffaceous, as the San Ignacio Formation seen to the north near San Ignacio. Locally within the unit are very shaley beds which interfinger with the Isidro. They are not actually mappable as separate units.

100 The highway climbs through the beds of the El Cien Formation and passes through the small settlement of El Cien.

On the left side is the 300 meter section at Cerro Colorado which is at the core of a major anticline. Most of the exposures on Cerro Colorado are the Oligo-Miocene marine El Cien Formation with a nonmarine tuffaceous redbed near the top. The tuffaceous redbeds and a conglomerate bed on Cerro Colorado can be traced across the peninsula into the "Comondu" Formation on the Gulf. Redbeds and fluvial sedimentary rocks become thicker to the east; where as, the marine Miocene rocks at the base of the hills become thinner. This indicates an eastern source for the volcanics. In the distance to the right the "Comondu" Formation can be seen on the south limb of the anticline. Progressively younger beds of the Comondu Formation are exposed to the right.

The name **"Comondu Formation"** (Heim, 1922) has been applied to volcanic and volcaniclastic rocks of Miocene to recent age in Baja California. Radiometric dating provides a basis for determining the relative positioning of the strata from widely separated areas.

The **Isidro Formation** was defined by Heim (1922) as exposures of greenish, whitish, and yellowish sandstones with interbedded greenish shales in the La Purisima area. The Isidro Formation is Early Miocene and overlain by a 22 million year old tuff from the Cerro Colorado area.

The widespread **Monterey Formation**, (now renamed the El Cien Formation) and San Gregorio Formations, were also defined by Heim (1922) for exposures of hard, clear, siliceous, diatomaceous yellowish shales in the La Purisima area west of Loreto and in the Magdalena Plain.

97 The highway passes through road-cut exposures of the El Cien Formation for the next three kilometers. Abundant splintered pieces of siliceous shale scattered over the surface indicate the presence of this formation.

92 The Magdalena Plain desert vegetation continues to be dominated by species of Jumping Cholla, Teddy Bear Cholla, Cardon, Elephant Trees, Creosote Bush, Pitaya Dulce, and Pitaya Agria. The Wash Woodland vegetation continues to be predominated by scattered specimens of the leguminous Acacia trees and Palo Verde. Another interesting and vicious looking plant of this region is the Creeping Devil *(Machaerocereus eruca).*

THE CREEPING DEVIL is a prostrately growing cactus because of the weight of the heavy stems. This sharp spined, tire menacing cactus sends out a dense network of branches which often produce adventitious roots where they touch the ground. As the prostrate branches slowly grow forward, the older hind parts die; this plant spreads and multiplies itself by asexual, vegetative reproduction. This mechanism of reproduction is often used by species which inhabit arid environments as a way to avoid the "water expensive" process of sexual reproduction which involves the production of moist flowers and seeds which require water for germination. Creeping devil grows in alluvial soils over the Llanos de Magdalena.

89 The chevron folds exposed in this roadcut are in the Monterey Formation.

FOLD IN MONTEREY FORMATION

88.6 The cliffs to the left offer a cross-section of the geology of this part of the peninsula. The upper thick pink tuffs and gray-green fluvial sandstones are the Comondu Formation; the middle yellow-gray sandy San Ignacio Formation is visible on the east side. It grades downward into the yellow, marine sedimentary rocks of the Monterey and Isidro Formations on the west side.

87.5 The highway passes through typical beds of the El Cien Formation.

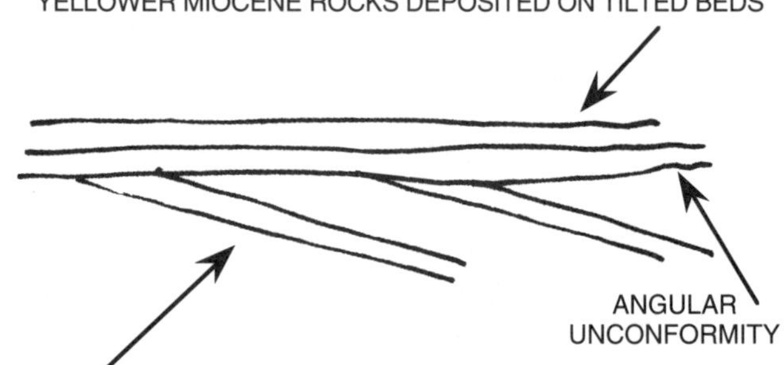

85.5 The contact between the brownish marine Paleocene Tepetate Formation and the overlying yellowish marine Oligocene-Miocene El Cien Formation is exposed on the hill to the left of the highway. The flat-lying contact between the formations is a slight angular unconformity. The rocks of the Eocene part of the Tepetate Formation are tilted and partially removed during a period of uplift and erosion producing the unconformity.

84 The highway passes through exposures of redbeds and yellow-brown sandstones of the Tepetate Formation. The roadcut above the highway has yielded a discocyclinid Coquina. Discocyclinids are very large marine foraminifera that grew to one centimeter in size.

83 The Pacific Ocean can be seen to the right. Some of the distant small mesas are composed of Pliocene Salada Formation which overlie the yellow-brown Tepetate Formation.

TEPETATE FORMATION

79 The hills to the left contain excellent exposures of the Tepetate Formation. The highway descends into an amphitheater-like area with exposures of the flat well-bedded strata of the Tepetate Formation. To the north is the brown Paleocene marine Tepetate Formation and the overlying yellowish El Cien Formation.

78.5 Along the highway the surface of the hills are littered with small flat shale fragments called "tepetates" (slabs).

77 The highway enters Arroyo Conejo. There are excellent exposures of the Tepetate Formation on both sides of the highway. It is worth a walk both up and down the east bank of the arroyo to view this formation.

Knappe (1974) stated that in Arroyo Conejo, "the **Tepetate Formation** consists of a series of Paleocene sandstones and shales exposed along the Pacific coast of Baja California Sur, Mexico. Foraminifera collected from a section of this formation in the vicinity of Arroyo Conejo are Eocene in age and correlate with benthonic and planktonic assemblages found in California, Oregon, Washington, and elsewhere. The oldest and youngest benthonic fauna examined correspond to the Penutian and Ulatisian stages of Mallory (1959). Planktonic species from the Tepetate section correspond to the *Morozovella aragonensis, Subbotina senni,* and *Truncorotaloides densus* zones of Schmidt (1970) which are approximately equivalent to the *Globorotalia formosa, Globorotalia aragonensis, Globorotalia palmerae,* and *Hantkenina aragonensis* zones of Bolli (1957, 1966).

Planktonic foraminiferal correlations suggest that the Ulatisian stage is lower and basal middle Eocene, and the Penutian stage is middle lower Eocene rather than middle Eocene and lower Eocene as suggested by Mallory (1959).

The foraminiferal assemblage identified from the section is characteristic of upper to middle slope deposits and includes (after Sliter and Baker, 1972; Bird, 1967): *Anomalinids, Bulimina* - costate forms, *Bathysiphon, Cibicidoides* - compressed forms, *Gyroidina* - rounded margin, *Siphogenerina* - costate forms, *Trifarina* - costate-spinose forms, and *Osangularia*."

76.2 Rancho San Agustin is located on a river terrace next to the arroyo.

75.4 The highway begins the ascent of a grade through the sandstones and shales of the Tepetate Formation.

74.5 To the right of the highway, a road-metal quarry exposes Tepetate sandstones and shales with more discocyclinid forams.

70.5 The El Cien Formation is exposed in this roadcut. The strata along the highway are near the contact between the tuffaceous fluvial Miocene Comondu sedimentary rocks and marine rocks. In the canyon to the left are exposures of the light-yellow to gray marine Miocene rocks. In the near distance are the overlying red tuffs and gray fluvial sedimentary rocks of the Comondu Formation. The El Cien Formation is represented by well-bedded siliceous and tuffaceous sedimentary rocks which include shales, siltstones and cherts. The marine Miocene rocks on the west side of the peninsula are stratigraphically correlatable with nonmarine fluviatile and tuffaceous beds of the "Comondu" along the Gulf of California.

69 The vista is to the southwest of the Pacific Ocean and the low mesas underlain by the Salada Fm. The canyons along the highway contain abundant exposures of light greenish to yellowish gray rocks of the marine El Cien Formation.

67.5 There is a fault in the roadcut at Rancho Aguajito.

The highway travels along the tilted erosion surface of the gentle Pacific slope and passes through the Miocene fluviatile volcaniclastic sandstones of the Comondu. The Highway follows this surface to Km 34 on the Gulf escarpment. In this stretch all of the roadcuts are in the fluviatile sedimentary rocks of the Comondu Formation which was nicknamed the "gray-green grunge" by a geologist who worked in this region

62 The dense dark evergreen trees with gray trunks scattered in this area are Palo San Juan (*Forchammeria watsonii*). It is a member of the Caper family (Capparidaeae) which produces an edible deep-purple fruit. Occasional scattered Datilillo are seen.

42 The modified badlands topography along this section of the highway is developed on the fluvial volcaniclastic sedimentary rocks.

The vegetation consists of Lomboy, Elephant Trees, Datilillo, Greasewood, Pitaya Agria, Pitaya Dulce, Cardon, Hedgehog Cactus, Palo Adan, Palo Blanco, *Acacias*, Jumping Cholla, Teddy Bear Cholla, and occasional Mistletoe.

It is not unusual along this stretch of the Highway to see Turkey Vultures (*Cathartes aura*) roosting in the Cardon Cactus and other trees.

Turkey Vultures (buzzards) or Zopilote de Cabeza Roja are ubiquitous throughout Baja and are protected by law for their value as "cleaner-uppers" of dead bodies. Turkey Vultures are naked headed, diurnal carrion eaters which are commonly seen scavenging on the ground in open country and along roadsides. Along the Highways, one may see a small group of them devouring a dead animal which they had detected by sight or smell while soaring high in Baja's skies. In flight, they appear to rock from side to side, and seldom flap their wings. They are also often seen perching on the tips of Cardon branches "sunning" themselves.

TURKEY VULTURE "SUNNING"

"Sunning" is a phenomenon not well understood. At one time it was thought that "sunning" was performed to warm the birds, but Turkey Vultures have been sighted "sunning" themselves in the coldest rains, winds, and on the hottest days of the year.

The most identifiable field characteristic of the buzzards is the way they soar in wide circles, holding their black wings in a broad "V" while rocking quickly from side to side.

In the southern part of the peninsula, Vultures and the black and white crested Caracara are occasionally seen feeding together on the same carcass.

36.1 The highway passes Microondas Matape.

34 on a clear day the view from the top of the grade is impressive. The viewpoint is located on the south limb of a major anticline in the Miocene rocks which dips south under Llanos de Todos Santos. To the northeast are Islas Partida and Espirito Santo; they are composed of highly faulted and tilted volcanic rocks of the Comondu Formation. To the east the rugged granitic spine of the Sierra de la Victoria (65-75 million years) begins on the peninsula east of La Paz and stretches south to the Cape Region. This range is separated from Llanos de Todos Santos by a major fault zone (Normark and Curray, 1968).

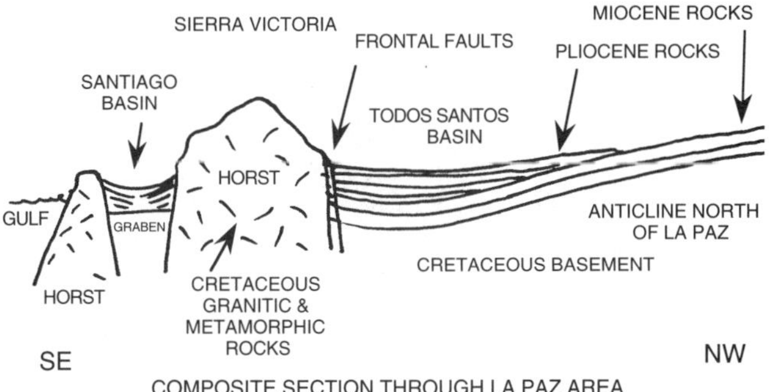

COMPOSITE SECTION THROUGH LA PAZ AREA

33.5 The highway descends through volcaniclastic sandstones of the Comondu Formation.

28 Due to overgrazing, Cholla Cactus has become a prominent element of the understory vegetation in this area. Scattered specimens of Pitaya Agria, Pitaya Dulce, Cardon, Acacia, Elephant Tree, Lomboy, Palo Adan, and Mistletoe are growing on the hillsides. The dark green shrubby Creosote Bush is the dominant plant of the Wash Woodland flora.

COMONDU FORMATION ON GULF

25 The view to the left rear from a low rise reveals the pink tuffs and gray-green fluviatile sedimentary rocks in the south limb of the San Juan anticline which rises to the north and dips out of sight to the south. The highway crosses the anticline in fluvial volcaniclastic sandstones above the highest pink tuff. These rocks were deposited prior to the tectonic opening of the Gulf.

The vegetation in this area consists of Lomboy, Palo Blanco, Palo Adan, Mesquite, Elephant Tree, Cardon, Pitaya Agria, Pitaya Dulce, Cholla, and Agave. This area is an ecotone (*See* 2:118.5) between the flora of the Magdalena Plain subdivision of the Desert Phytogeographic Region and the Cape Phytogeographic Region.

The **CAPE PHYTOGEOGRAPHIC REGION:** The Cape Region and the California Region are outside of the Desert Region which covers most of the peninsula. The Cape Region includes the Cape mountains and part of the Sierra de la Giganta. This region has the highest rainfall on the peninsula. Most precipitation in the Cape Region occurs during the tropical summer storms.

The vegetation of the Cape Region is divided into two areas, the **Oak-Pinon Woodland** and the "impoverished" **Arid Tropical Forest**. The vegetation at lower elevations is thorn scrub with elements of tropical deciduous woodland. At higher elevations the vegetation is chaparral-like, and then evergreen Oak and Pinon Pine forests. Floristically, the Cape Region resembles the Pacific coast of southern Mexico.

The **Oak-Pinon Woodland** community occurs in the granitic soils of Sierra la Victoria to an elevation over 6,000 feet (the transition life zone). The dominant species are Oaks, Black Oak and Pines, Palmita, Laurel Sumac, and Madrona.

The **Arid Tropical Forest** of the Cape Region is an "impoverished tropical jungle". The trees of this forest are represented by the leguminous trees Palo Blanco and Palo Mauto, Mesquite, Acacia, Coral Tree, the edible Plum Tree; Palo Verde; and Plumeria with an understory of Palo de Arco, Pitaya Dulce, Lomboy, and Palo Adan.

The avifauna of this region is not very diverse, but the following species are commonly seen:

BIRD NAME	LIKELY LOCATION
American Kestrel	Wires and fence posts
American Robin	Common in woodland openings, forest borders
Belding's Yellowthroat	Stays low in grassy fields, shrubs, marshes
Blue-Gray Gnat Catcher	Perched on vegetation
California Quail	On the ground
Common Ground Dove	On the ground
Costa's Hummingbird	On red or yellow flowers
Crested Caracara	Feeding on carrion
Gila Woodpecker	In the scrub
Gray Thrasher	Looking in ground litter for food
Ladder-Back Woodpecker	Flitting in the air
Loggerhead Shrike	Perched on wires, fences
Northern Cardinal	Woodlands, streamside
Pyrrhuloxia	Common in thorny bushes and mesquite
Turkey Vultures	Soaring in sky looking for carrion
Varied Bunting	Perching on vegetation
Western Meadowlark	Fence posts and wires
White-winged Dove	Dense mesquite, riparian woodlands, open desert
Xantu's Hummingbird	Feeding at tubular flowers
Yellow-billed Cuckoo	Flying through vegetation
Yellow-eyed Junco	Coniferous and oak slopes
Yellow Warbler	Wet areas, woodlands

Peregrine Falcon or Halcón peregrino is distinguished by a "helmet" formed from a black head, neck, and wedge extending below the eye. This large sized bird flies fast and rarely soars. Peregrine falcons inhabit open wetlands near cliffs and are regularly seen at coastal bays and lagoons in winter. Occasionally, they are seen in cities, on bridges, and tall buildings. They mainly prey on ducks, shorebirds, and seabirds.

Mexican Chickadees, White-winged Doves, Cactus Wrens, Gilla Woodpeckers and Falcons are commonly seen in this area.

18.2 There is an excellent view of La Paz, the gulf coast capital of Baja California Sur. La Paz is considered to be one of the best sport fishing locales in the world. The sport fish of the area include Marlin, Roosterfish, Mahi-Mahi, Cabrilla, Yellowtail, Sierra, and many others.

17 The side road leads to the phosphate mining settlement of San Juan de la Costa. Phosphates are mined and shipped elsewhere in Mexico for use as a fertilizer. This mine supplies about 40% of Mexico's phosphate needs.

About 90% of the world's production of phosphorous is from sedimentary phosphate rock (phosphorite) of marine origin such as that found and mined in this area.

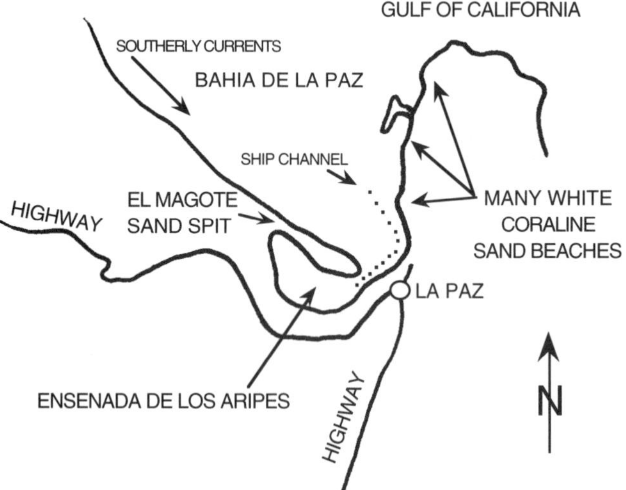

15 The highway passes through the small bayshore area of El Centenario.

Bahia De La Paz, the largest bay on the peninsula's east coast, and the muddy sand spit of El Magote are visible on the left. El Magote is a narrow sand spit formed by the southward transport of sand by currents along the eastern coast of the peninsula. It separates the main body of the bay from Ensenada de Los Aripes, the shallow inlet to the west

13 The highway follows the bay across a supertidal flat with numerous mangroves.

8.9 This turnoff leads to the right to the La Paz International Airport.

6.8 The road to the right is the most direct route to the cape. Proceed to Padre Kino, right to Forjadores which becomes the main highway to the cape, or go past Padre Kino about 5 blocks to a major shopping area. There are at least two signed streets in town which direct traffic toward Cabo San Lucas.

To reach Pichilingue and the Ferry docks, continue to the left on Abasolo to the Malacon.

0 **LA PAZ**: The Malacon at La Paz is a coconut palm-lined, sea wall and pedestrian walkway combination which meanders along the bay in the main part of La Paz. At the intersection of 5 de Febrero, Highway 1 becomes Mexico 11 and quickly turns to the north along the bay toward Pichilingue. It is well worth the side trip to Pichilingue to see the beautiful coves and biological sand beaches. (See the cover of the book.)

Stop at Museo Antropologico de Baja California Sur at Ignacio Altamirano and 5 de Mayo. The museum features good geological, historical, and anthropological exhibits.

SIDE TRIP TO PICHILINGUE AND THE FERRY DOCK VIA MEXICO 11:

Mexico 11 to Pichilingue wraps around Ensenada de Los Aripes and leaves by way of Paseo Alvaro Obregon. There are a number of groins built along the beach to slow the drifting of sand due to currents which flow into the bay.

1.5 This road travels primarily through the pink tuffs and the gray-green volcaniclastic sedimentary rocks, massive fluviatile bouldery conglomerates, and mega-breccias of the Comondu Formation.

2 This is Hotel Palmira. The salt marsh estuary next to this hotel used to be a national park. It is now largely filled by natural sediment. As you travel in the cape region, watch for herons, egrets, storks and other marsh birds.

The **Cape Phytogeographic Region** is vegetated by a dense thorny, Acacia dominated flora known as thorn scrub. Thorn scrub grows where summer precipitation exceeds 300 mm/Yr.

2.5 The highway becomes a divided road.

2.8 The highway climbs a small grade through faulted pink tuffs and fluvial volcaniclastic sandstones of the Miocene Comondu Formation. This is the best place to observe exposures of these rocks which are exposed for over a hundred miles in the cliffs along the Gulf northwest of La Paz.

Mexico 11 will cross several steeply dipping north-south normal faults related to the frontal faulting of the Sierra Victoria.

4.1 On the inbound lanes the highway passes the coralline biological "sand" swimming beaches of Coromuel. These beaches have been developed into a public plaza and bathing pavilion. A native wild fig tree (*Ficus palmeri*) is across from Coromuel beach (*See* 7:98).

Rainfall, scarce in La Paz, arrives infrequently in the form of violent tropical storms called Chabascos. As a result, summers are hot and dry. The name "Coromuel" refers to the cooling wind which comes from the cooler Pacific Ocean to the south across the peninsula and refreshes every summer afternoon.

4.5 The ship channel markings delineate a narrow channel very close to the near shoreline. The southward movement of sand along the Gulf has formed El Magote and piled sand in the mouth of the Ensenada. Tidal currents scour the channel as water moves in and out of the Ensenada.

5.1 The Las Conchas Resort is located at El Caimancito. The small offshore rock, Islita Caimancito, received its name because it resembles a lurking alligator (Caiman).

The highway passes through more volcaniclastic rocks and tuffs of the Miocene Comondu Formation which is exposed around the resort. On a clear day the major San Juan anticline in the Comondu Formation can be seen on the west side of Bahia de la Paz. This anticline and one to the north stretch for over 120 miles north of La Paz. The same bed can be traced for nearly the entire distance. These anticlines are strong petroleum prospects, since the oil-bearing Cretaceous geosyncline underlies the volcanics to the eastern edge of the peninsula.

5.3 The divided highway ends and returns to a single lane in each direction. There are exposures of pink tuffs for the next several kilometers.

Because of unfavorable edaphic (soil) features, the vegetation is very sparse on the surrounding volcaniclastic tuffs and consists of sparse Cardon, Lomboy, Palo Adan, *Acacia*, and Candelabra.

8.2 Punta Prieta (dark point) is the location of the government owned and operated Pemex oil refinery docks and storage tanks. Oil is transhipped from mainland Mexico across the Gulf to Baja for storage and distribution. Even though oil bearing strata exist here, there is no developed source of oil in Baja California.

9 A good exposure of the coarse conglomerates of the Comondu Formation can be seen here.

9.4 The view into a small bay with a Mangrove swamp and a small estuary reveals the first of a series of shallow coralline biological "sand" coves with azure blue water formed by recent submergence of this area.

10 A small restaurant is located on the shore of this beautiful playa. The beach is formed by a light-colored coralline biological "sand" and backed locally by a mangrove swamp.

11.8 As the highway skirts another picturesque cove, there are excellent exposures of very large mega-breccia fluviatile sedimentary rocks of the Comondu Formation.

14 The highway cuts through the large mega-breccia fluviatile sedimentary rocks of the Comondu Formation.

The water in this beautiful cove is very blue-green in stark contrast to the light-colored coralline biological "sand" beach. Great Blue Herons are often seen in the Mangrove swamp located along the shore of the cove.

The Great Blue Heron or Garza Azul is often seen in the marshes, on the beaches, or in the fields of Baja. Its tall lean solitary figure can be seen standing motionless in a pool of water or advancing slowly one step at a time lifting each foot stealthily from the shallows without a ripple.

Herons may stand still as a statue for over half an hour while waiting for prey to come. With a lightning-quick forward lunge of their long neck and bill, the heron captures its prey. It prefers fish but will also eat birds, small mammals, insects, snakes, frogs, and crustaceans.

Whether on land or in the air, a Great Blue is recognized by the long snakelike neck held in an "S" shaped curve, the slow wing strokes, long legs, and nearly two meter wing span.

14.5 The highway skirts a cove through a Mangrove swamp and traverses the super-tidal flat inland from it. The vegetation on the supertidal flats is sparse and impoverished due to the highly saline soil (*See* 16:35). A few Cardons are seen on the hillsides along with scattered specimens of Lomboy, Organ-pipe and Mission Cacti, and scattered Acacia trees. A small cardonal covers the high flat behind the lagoon.

15.4 The highway passes through a nearly vertical roadcut composed of massive resistant fluvial mega-breccias. Isla Pichilingue can be seen across the lagoon.

16 This is the Ciencias Del Mar, Departmento de Geologia Marina Labs and Research facilities branch of the Universidad Autonoma de Baja California, Sur (UABCS) located on the tidal flats.

16.5 The ferry dock with several loading areas is to the left. Many pleasure yachts are docked here.

At the ferry docks the rocks in the roadcut are andesite mega-breccias. Beyond the ferry dock the gravelly road to the right leads to the water and a beautiful public beach, Playa Publica Pichilingue, situated in a beautiful cove. This is a good place to explore the coast.

The highway crosses a causeway built to reach Isla Pichilingue. It is now referred to as Peninsula San Juan Nepomuceno on some maps of La Paz.

17.2 Bahia Pichilingue is a deep-water bay used as the main shipping point for La Paz Harbor. It avoids the narrow and difficult passage into the main harbor of La Paz.

BAJA'S FIRST SETTLEMENT: Isla Pichilingue is the site of the first attempt to settle by Hernan Cortez the conqueror of Mexico, in 1535 in Baja California. However, supply problems resulted in the failure of that first settlement. Jesuits in 1720 established a mission here, but it also failed. La Paz finally became a permanent settlement in 1811 and became the state capital in 1829 after a hurricane destroyed the first capital, Loreto.

Go straight ahead to continue to Balandras (cover Photo) and Tecalote.

18 The road passes a coralline sand beach and climbs through the andesite breccias.

19 The road passes through a dense cardonal with Elephant Trees and Mesquite.

22 Junction

The right fork goes 2 Kilometers. to the sandy beaches of Tecalote.

The left fork goes 1 Km. to Balandras Bay which is one of the most beautiful bays in Baja. The cover Photo was taken from the ridge on the other side of the bay.

You can wade to the right around the small rocky point, (it's shallow) walk the beach to the next small point, and then wade to the end of that point to see the Sombrilla [umbrella].

This is a good place to view a solution-cut notch and terrace. A solution cut terrace is produced in arid areas where there is not much moisture and weathering of the hillsides is very slow. The rocks along the shoreline are moistened by salt spray. The increased moisture promotes increased weathering and the notch. Another effect which aids in the development is the salt in the water which crystallizes in the pore spaces and cracks and pries the rock apart just above the water line

SOMBRILLA AT BALANDRAS

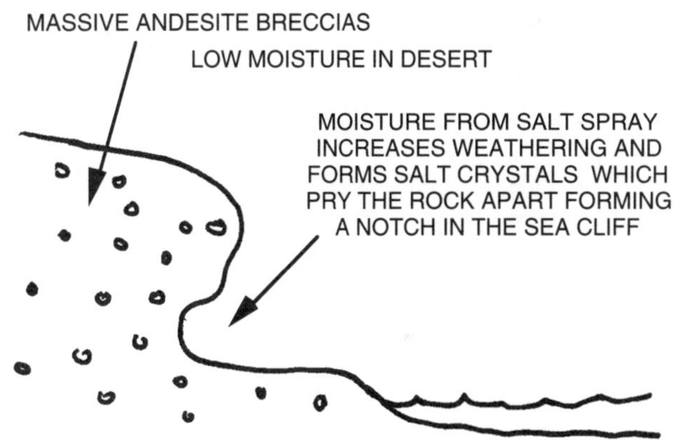

Where To Now? Return to La Paz and continue to the south on Highway 1 using the La Paz to Cabo San Lucas log for your guide.

Log 9 - La Paz to Cabo San Lucas [211 kms = 131 miles]

The Highway follows the Gulf fault zone in an alluviated graben-syncline. It then turns to climb over uplifted granitic and metamorphic horsts, first at El Triunfo, then San Antonio and San Bartolo to descend to the Gulf at Buena Vista. The highway then travels south on marine and nonmarine sedimentary rocks, along the relatively flat Santiago Trough between two high granitic horsts, to San Jose de Cabo. Finally the highway skirts the southern edge of the Sea of Cortez on alluvial fans to Cabo San Lucas and its granitic headland.

221 The best route out of La Paz toward the Cape is to go south on Padre Kino, then diagonally right on Forjadores.

210.5 The divided highway ends. The campus of the Universidad Autonoma de Baja California Sur is on the right.

During several particularly violent storms the Baja Highway south of this point was severely damaged in 1982 and 1984.

210 Andesite and Andesite breccias of the Comondu Formation are exposed for the next several kilometers as they dip toward the highway from the left. The white bouldery granitic outcrops of the cape region form the high hills to the far left.

South of La Paz the edge of the Sierra la Victoria range is remarkably straight which is a result of the recency of faulting in this area. The highway follows the Todos Santos graben-syncline which is situated between the Sierra la Victoria Fault Zone and the south limb of the anticline in the tertiary sedimentary rocks. This graben is filled with Pliocene to Recent sediments.

The Mesozoic granitic and metamorphic rocks of the Sierra la Victoria are similar in age, relationships and rock type to the Peninsular Range Batholith. They were all part of the same granitic belt prior to the Cenozoic extension which separated Baja from the mainland. Age relationships in the Gulf indicate that this separation took place at the mouth of the Gulf of California about 4-5 million years ago and continues today.

186 The highway passes through the Cape Phytogeographic Region of Baja. The predominant plants of the Cape Region flora in this area consist of Cardon cactus and a mixed "forest" of leguminous trees such as Palo Mauto, Palo Blanco, Acacia, Screwbean Mesquite, and Palo Verde. Along the highway the dominant plant of the Cape Region appears to be the Cardon cactus. Next in order of decreasing dominance are Elephant Trees, Lomboy, and Palo Adan. Other less dominant plants seen growing in the Cape Region are Jumping Cholla (primarily in disturbed soils), Plumeria, Organ Pipe Cactus, Creosote Bush, and the yellow-flowered vine-Yuca. Yuca is a morning glory relative which covers many plants like a large net.

THE BEAN TREES OF THE CAPE REGION: Trees are a conspicuous and important part of the desert flora of Baja. Many of Baja's desert trees are legumes which produce a "bean" pod containing numerous "beans" (seeds). Leguminous trees, members of the family Mimosaceae, occur most often at low elevations in the desert, arroyos, and foothills below 1,000 meters. They are common and conspicuous in the Cape Region. The most commonly seen leguminous trees in the Cape Region are Vinorama (*Acacia brandegeana*), Palo Mauto (*Lysiloma divaricata*), Palo Blanco (*Lysiloma candida*), and Mesquite (*Prosopis pubescens*). The following table will help identify these four common roadside legumes of the Cape Region:

PALO MAUTO

Another non-leguminous tree commonly seen growing with the four leguminous tree species of the Cape Region is Palo Verde in the Senna family. Species of the Palo Verde, a conspicuous and characteristic tree of the Sonoran Desert

	Vinorama *Acacia brandegeana*	**Palo Blanco** *Lysiloma candida*	**Palo Mauto** *Lysiloma divaricata*	**Mesquite** *Prosopis pubescens*
Bark		Silvery-white,	Grayish-brown smooth	Thin, flaky
Spines	Slender, straight occurring singly	One of the few Baja legumes without thorns	None	Short, awl-shaped spines in pairs at Nodes
Flower	Yellow spikes	Creamy white ball-like clusters	——	Yellow spikes
Pods	Long and slender turn coppery red at maturity	Long, thin-walled	Twisted pods	Tightly spirally coiled pod - a "screwbean Mesquite"
Range	Open flood plains, lower arroyos, and mesas	below 600 meters	above 300 meters	below 1200 meters, near water sources

north of the U.S./Baja border, grows all the way to the Cape region. *Cercidium praecox* is the Palo Verde of the Cape Region. It is easily distinguished from the leguminous trees by its bright green photosynthetic trunk and branches (cladophylls) and its leafless appearance. Palo Verde may be mistaken for a leguminous tree since it produces a pod-like fruit which contains seeds about the shape and size of pod beans

185 This is the turnoff to Todos Santos. Just south of the turnoff are large areas which have been cleared of native vegetation for grazing cattle.

The Plumeria is also a common plant in this area.

THE PLUMERIA OF BAJA: Plumerias grow as a shrub or tree up to 9 meters tall in Baja's Cape Region, from just south of La Paz to Cabo San Lucas, along canyons, and foothill slopes. For most of the year Plumeria is leafless but is easily recognized by the showy long, white, tubular flowers clustered at the ends of branches. The white blooms are abundant after summer rains. Plumeria are frequently seen in the gardens of ranchos where they are planted as ornamentals.

178 Over a small rise is a distant view of the high massif of the Sierra la Victoria. The roadcuts are in pinkish alluvial material derived from the decomposed granitic basement.

176.8 Large Crested Caracara, the national bird of Mexico, are occasionally seen on the ground. Several other types of birds, including the red-tailed hawk, white-winged dove and Aplomado Falcon, are often seen sitting on the ground and on wires along this stretch of the Highway.

The Crested Caracara or Quelele of the open desert scrub is a long-legged scavenger in the falcon family which spends much of its time on the ground. In flight the larger head and beak, longer neck, white throat and black and white-banded tail set it apart from the vultures. Its red-orange, bare facial skin is easily identifiable.

One of the best times to see the Caracara is on the highway, in the early morning, as it is the early bird that gets the road kill. Because of its long legs, the Crested Caracara most often walks along the ground gathering food which consists of small creatures, carrion, and sometimes plant matter. They are a dominant bird often seen stealing food from other birds.

175 As the highway crosses a series of undulating shallow ravines, it passes over the Sierra la Victoria Fault and into the main mass of the Sierra la Victoria which have been uplifted along this fault. Along the highway south of La Paz, the frontal fault of the Sierra la Victoria cuts off the metamorphic rocks on the edge of the batholith. The highway crosses directly from the alluviated plain across the frontal fault into granitic rocks with very few metamorphic rocks.

174 The highway passes through roadcuts of pinkish granitic rocks.

171.5 The highway descends along a rather large arroyo and passes through alluvial fan gravels derived from the Sierra la Victoria.

168.4 Phyllitic schists are exposed here.

168 The highway ascends through the foothills of Sierra la Victoria. There is a small mining area with a mine shaft and mine workings on the right side of the highway. These foothills are pocked with prospectors' holes. The hills are covered with Palo de Arco (*Tecoma stans*).

PALO DE ARCO is a small shrub or tree which produces large golden-yellow flowers after rains. The stems of this shrub are used to make shipping crates for raw sugar (panocha) produced at the sugar cane mills of Todos Santos.

PALO DE ARCO

167 Altered and brecciated rocks are visible in this roadcut. A fault zone separates the gneisses from the weathered granitic rocks.

165.2 View of the old mining region of El Triunfo.

VIEW OF EL TRIUNFO

164 The small mountain village of El Triunfo was established in 1862 with the discovery of silver and gold. This town was the center of a major silver mining operation after 1862. Two yellow church towers rise between the smoke stack and mine buildings on the hill where $50,000 worth of silver and gold were produced monthly. These ores are concentrated in dikes in the

metamorphic rocks near the contact with the granitic rocks. A side trip to the main mines can be taken on the first road to the right after you enter town and cross the bridge. Follow it for two blocks, go left at the wash, and cross an old brick bridge. The road wanders to the smelter and the old tower. It is possible to climb inside the tower and look up the old brick smoke stack at a small patch of sky.

160 The highway passes through outcrops of mixed gneissic and granitic rocks.

The hillside vegetation is primarily an impoverished rain forest of several species of leguminous trees. Other plants seen are Pitaya Dulce (it looks almost like a Cardon in this environment), Elephant Tree, and Palo Adan. At this southern more mesic latitude, Palo Adan, a relative of Ocotillo, becomes less abundant and is largely restricted to the lower sandy Wash Woodlands.

159.4 There are prominent granitic dikes in the roadcut. The rocks at the crest of this pass are gneiss and dark-colored granitic rocks invaded by numerous granitic dikes.

SMELTER SMOKE STACK

157.5 The highway descends into the town of San Antonio, another “old time” mining town.

A number of dikes and sills cut the metamorphic and granitic rocks in this roadcut.

156 The historical gold and silver mining town of San Antonio is located in the foothills of the Sierra la Victoria. This mining town was founded in 1756 when Gaspar Pison opened a silver mine over 100 years before the establishment of El Triunfo. Now San Antonio is a cattle ranching center.

155.5 The highway ascends a grade with a view of Llanos San Juan de Los Planes, which extends to the north to Punta Colorado and to the Gulf. The volcanic and granitic Isla Cerralvo is visible offshore. The San Juan plains are part of the alluviated graben into which the highway will descend after cresting the grade.

IMPOVERISHED RAINFOREST

LOMBOY- CARDON- PALO BLANCO- PITAYA DULCE

The vines entwined on many of the trees of this area are either San Miguel (*Antigonon leptopus*) or Yuca (*Mersemia aurea*).

SAN MIGUEL VINE: The vine-like plant which grows on some of the trees of this region is known as San Miguel or Coral vine. This member of the buckwheat family reminds one of the commonly cultivated bougainvillea of Southern California. Its seeds and potato-like tubers (modified underground stems) were eaten by Indians.

YUCA

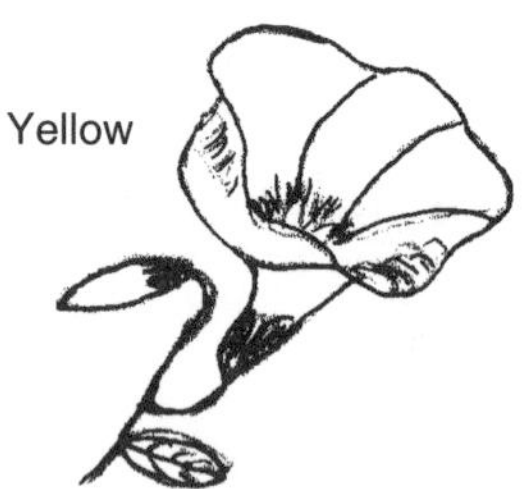

The **YUCA** vine is a member of the morning glory family which climbs upon and covers many of the plants of this area. When the Yuca die, they cause the host to appear as if it is covered by a net. Yuca have bright yellow trumpet shaped flowers that bloom all year, especially after a rain.

155 Gneisses are exposed in the roadcut. The turnoff provides a good view of San Antonio.

154 This is the best view of the alluviated Llanos de San Juan region.

153.5 The highway proceeds through metamorphic rocks cut by prominent granitic dikes while crossing over the main horst and the high granitic spine of the Sierra la Victoria. The highway drops into a graben which contains the towns of Santiago and Miraflores before crossing another smaller horst and descending into the Gulf town of San Jose del Cabo.

151 The Mimosas in this area seems to be primarily restricted to the well-drained higher slopes of the Sierra la Victoria.

The dominant vegetation of the Wash Woodlands is Palo Adan, Elephant Tree, Jumping Cholla, Organ Pipe Cactus, Pitaya Agria, Acacia, Pitaya Dulce and Lomboy. Abundant Cardons are growing in a narrow belt at the base of the hills.

149 Road to Rancho Los Encinos.

Northern Mockingbirds or Centzontle Norteño are commonly seen in this flat region. They are robin-sized birds with white breasts and black wings which show white patches in flight and a long narrow tail. Mockingbirds are expert mimics and repeat most of the songs of local birds while perched or in flight. These mimics sing more at night than do other members of the family Mimidae, an interesting but irritating fact if you are trying to sleep!

147.5 The highway splits in a wide "Y". To the right is San Antonio de la Sierra. The fork to the left continues toward Cabo San Lucas.

The outcrops are in granitic and metamorphic rocks with numerous light-colored dikes.

145.7 The granitic rocks at this kilometer mark have been dated at 73 million years. The flora is dominated by Mimosa and Pitaya Dulce. Palo Adan is seen as the highway climbs the slope.

Throughout this region, the Loggerhead Shrike may be seen.

Loggerhead Shrikes or Cabezón are easily recognized by their heavy hooked beaks, broad black masks, and large white wing patches which contrast with their dark wings. Its head and back are bluish-gray and their underparts are white. They are known as "butcher birds", since they habitually capture small lizards, insects, rodents, and small birds and impale them on cactus thorns, tree thorns, or barbed wire. This prevents their meal from escaping and stores it for later.

145 The forests of the Cape region are "double-canopy" forests. Palo Mauto and Acacia form a higher canopy which shades the lower plants; this produces a moister more mesic environment and reduces evaporation. The taller Organ Pipe cactus grow at the bottoms of the slopes where water collects as it drains off the surrounding hills.

143 Palo Mauto and Plumeria now occupy just the very high parts of the hills. *Acacia* begins to take over and becomes the dominant canopy tree.

137 After Rancho El Rodeo the Cochal or Candelabra Cactus is seen to the right of the highway. A few Cardon are mixed with the Organ Pipes along this stretch of the highway.

THE FRUIT OF THE SECOND HARVEST is the large tasty red fruit of the Pitaya Dulce (*Lemaireocereus thurberi*) which ripen in the late summer and fall. Gulf coast native Indians would gorge themselves on these fruits when they were in season during the "first harvest". The fruits, like raspberries, contain small black seeds (achenes) too numerous to remove. The Indians knew that the "first harvest" would only last a short time and then they would be hungry again. In an effort to utilize all of the resources available, the Indians defecated in one spot and would later sift through the dried feces and separate the small seeds of the Pitaya Dulce, the "fruit of the second harvest". The seeds were ground into meal and made into the "bread of the second harvest".

136 Weathered granitic rocks are cut by dikes in this roadcut.

134 The highway begins to roughly parallel an arroyo vegetated primarily by Cheese Bush.

131 There are exposures of light-colored granitic rocks with darker zenoliths along the road.

A wild Fig tree is growing here and Yuca are becoming quite common.

130 A small rancho is located on the left. Broom Baccharis grow in the disturbed soils along the side of the highway with nearly leafless branches crowned by a green, stiff, broom-like mass. Often bundles of branches are tied together and used as a broom.

128.5 Growing in profusion around the tropical agricultural village of San Bartolo are avocados, sugar cane, lemons, limes, mangos, papayas, figs, date palms and fan palms. The vegetation of this arroyo is lush and tropical with *Acacia* as the predominating plant. Occasional Palo Adan, Elephant Tree, Palo Mauto, and Organ Pipes are scattered among the Acacia.

South of San Bartolo the highway follows the northeast bank of the arroyo. Exposures of granitic rocks occur in the arroyo and on the hills.

125 The stream terrace on the other side of the canyon was formed when sea level was higher and the streams graded the fans to a higher level.

124 Massive exposures of fresh granitic rocks are exposed just before Puente El Saldito.

123 The highway drops almost into the arroyo where roadcuts expose the stream terrace material and then climbs up onto the terrace surface

120 This area was a basin at one time which in-filled with very coarse granitic debris. Now this basin is being uplifted and dissected and the hills are covered with the gray granitic gravels that underlie this uplifted area. These gray granitic rocks coupled with the sparse impoverished vegetation give the impression of a burned over area. However, this impoverished forest is lush and green during the summer rainy season.

119.5 The highway crosses the main wash of San Bartolo with outcrops of the sub-Recent to Pleistocene sediments. These sediments are uplifted alluvial fan material composed of granitic debris from the Sierra la Victoria.

Mimosa (*Lysiloma sp.*), Tamarisk, and Lomboy are growing abundantly along the edge of the wash. Frequent floods, indicated by the rarity of trees, occur in this wash. The high energy of the floods is indicated by the width and depth of the wash. Cheese Bush is the dominant shrub of the wash bottoms. It is able to reestablish itself more quickly than other plants; it can grow in the washes between the intermittent floods.

118 There are four levels of terraces which indicate different periods of uplift. The main older fan surface is on the far side of the yellow cliffs. Just below is a secondary surface which has large Palo Verde, Palo Mauto, and *Acacia* growing on it. A third level is closer to the highway with some smaller trees growing on it. And the fourth is the level of the wash bottom next to the highway. The higher hills beyond the fourth surface are alluvial gravels from a still older fan level.

113.5 The highway crosses Arroyo Buenos Aires and reveals good exposures of alluvial fan material with large granitic blocks and sandstone lenses. These blocks are nearly all granitic with only a few metamorphic clasts.

Some parts of the hills to the southeast of Los Barriles are metavolcanic rocks. Moreno dated the metamorphic rocks on the west side of the Sierra at 73 million years. The granitic rocks near El Triunfo have been dated at 73 million years and those nearer the Cape at 63 million years.

101 The highway passes through a relatively flat graben which has been filled with at least 1500 meters of nonmarine and marine sediments, has been uplifted, and is now being dissected. The two sides of the graben can be seen in the long line of granitic hills to the southeast and in the edge of the granitic spine of the Sierra la Victoria to the west.

This thick series of Neogene (Miocene to Recent) nonmarine and marine basin-filling sedimentary rocks has been named: 1) the Coyote Redbeds, a Miocene alluvial fan deposit; 2) the Trinidad Formation, a middle to upper Miocene beach to outer shelf member and a upper Miocene to Pliocene slope and basinal deposit member; 3) the Salada Formation, an upper Pliocene shoaling upward sequence of outer shelf to beach deposits; and 4) a Pleistocene alluvial fan Sequence (McCloy, 1984). Fossils are exposed in numerous arroyos in the basin.

92.5 Exposures of the marine Salada Formation with alternating fine and coarse-grained yellow sandstones can be seen in the roadcut.

The road to the right leads to Las Cuevas (the caves) where there are Indian pictographs (paintings on rocks) (*See* 5:59.5). The highway to the left leads 12 kms to La Ribera on the coast: then it goes 30 kilometers south along the Gulf to Cabo Pulmo, the only true coral reef in the Gulf of California.

Cabo Pulmo consists of a large bay with a small fishing village and a rustic campground. Fishing is excellent and skin diving is great along the reef with abundant tropical fish. Just beyond Cabo Pulmo is Punta Frailes, the eastern most point in Baja California.

CORAL REEFS are found where 1) the average water temperature is 78° and the coldest month is not below 60°, 2) there is average salinity, 3) there is a lack of turbidity and 4) sunlight (less than 150 m deep). Actually, corals can and do live in deep and cold water. These requirements are for the calcareous algae which bind the corals together to form the reef.

89 At the crest of the hill is a good view of the Santiago graben. The Neogene formations stretch for several miles to the base of the hills to the west. The fault at the base of the hills trends north-south, roughly parallel to the highway. The hills to the east are on the east side of the graben.

Caliente Manantials hot spring is located to the west along the fault near Agua Caliente.

88 Excellent exposures of the turbidites of the Salada Formation can be seen along the highway for the next km. They are massive sandstones with interbedded shales and siltstones.

VIEW TO WEST IN SANTIAGO GRABEN

VIEW TO EAST IN SANTIAGO GRABEN

84.7 This road leads to the community of Santiago. The Pericu Indians once inhabited this region and revolted against the Spaniards in 1734.

Spanish Jesuits first arrived in Baja at Loreto in 1697 where they established the first and oldest mission in the New World. The purpose of the mission was to convert the native Indians to Christianity. The Jesuits were the first farmers of the peninsula and worked very hard to develop an irrigation system that allowed them to grow date palms, locally known as Datil, figs and olives. The only labor available was the unskilled labor provided by the native Pericu Indians. The conditions of the peninsula were so harsh that the Jesuits were never able to raise enough food for themselves and all of the Pericu, so only a few Indians lived at the mission. The rest continued to live as they had before the arrival of the Spanish and came into the mission occasionally or for religious holidays.

The Jesuits were evicted from Baja by decree of Charles III of Spain. Before the Jesuits were evicted, the Pericu rebelled against the priests in 1734 and killed priests, mission workers, and even the crew of a Spanish Galleon that had docked at Santiago for provisions and water. They rebelled because they did not want to accept Christianity; they were polygamists and looked to powerful witch doctors for guidance. The Jesuits tried to stop polygamy by reprimanding the Pericu in public for this un-Christian life style and tried to discredit the witch doctors. After the rebellion the Pericu returned to living as they always had. In 1768 the Jesuits were evicted and replaced by the Franciscans, then the Dominicans. The Indians died out as a result of diseases spread by soldiers.

82.3 Crest over a small rise for a panoramic view of the graben. The straight nature of the fault of the Sierra la Victoria and the flat nature of the sedimentary rocks that filled the graben are obvious.

81.8 The Tropic of Cancer is the boundary between the Temperate Zone and the Tropic Zone.

70.9 The highway passes the turnoff to Miraflores. The craftsmen of Miraflores produce fine leather work.

The road into Miraflores offers a view of the massif of the Sierra la Victoria. There are two very high peaks on either side of a large pass almost straight ahead. The peak to the left is Cerro Picacho la Laguna, 1,892 meters. The peak to the right is 1,824 meters.

These peaks show scars which are evidence of debris or mudflows which have stripped the vegetation along their path.

67 The highway drops into an arroyo with fine-grained fanglomerate sedimentary rocks on both sides and coarse fanglomerates above.

53 On the flats the vegetation has thinned. The vegetation is locally dominated by Elephant Tree, Jumping Cholla, Lomboy, Candelabra cactus, Palo Mauto, Palo Zorrillo, and scattered Acacias, Pitaya Dulce, Cardon, Palo Verde, Fairy-duster Mimosa, and Mistletoe.

49.5 To the east are exposures of the sediments on the edge of the Santiago basin. The beds are tilted up toward the edge of the horst.

43 This turnoff leads to the Los Cabos International Airport. Its a four lane highway to the cape.

35 Near San Jose del Cabo the Lomboy, covered with a parasitic Mistletoe, becomes the dominant plant of the area (more than 2/3 of the vegetation). Acacia and Cholla comprise most of the rest of the vegetation.

33 The highway forks here. The right fork continues to Cabo San Lucas. The left fork goes into San Jose del Cabo. There are interesting orthogonal joints in the granitic rocks across the street from the Pemex Station. Since there are abundant granitic rocks in this area, the beautiful white beaches are all composed of quartz and feldspar.

27.8 A cardonal of Cardons is growing in the arroyo along the beach. On the hill there are Cardons, Pitaya Dulce, Palo Adan, Palo Mauto, Elephant Trees, Cholla, Pitaya Agria, and Lomboy.

North of the village of Palmilia, Lomboy, Acacia, Elephant Trees, some small Pitaya Dulce, Pitaya Agria, and Mimosa (Lysiloma sp.) dominate the flat areas. In the washes Palo Adan predominates. Most of the vegetation is no more than 2-3 meters tall which is characteristic of the short thorn scrub of the Cape Region.

27 The granitic rocks are overlain by fluvial sedimentary rocks (mostly decomposed granitic rocks). These exposures continue for a number of kilometers down the coast.

22 Sweeping views toward the cape are visible along the highway for the next several kilometers. The surf is much stronger here due to the proximity of the Pacific.

21.1 The highly weathered and jointed pink granitic rocks exposed in this roadcut have been dated at 63 million years.

19 The highway continues to traverse the dissected undulating alluviated fan surface.

16.4 There are a number of relatively unweathered and well-jointed dikes which cut the weathered granitic rocks.

15 The palm oasis of Hotel Cabo San Lucas is nestled along the coast. Many of the small offshore rocks and points along this coastline support small coral colonies.

11 Two fig trees are on the low hill to the left.

CABO SAN LUCAS

6 There is a spectacular view of the cape, the bay, the arches and stacks at the point of the Cape.

FRIAR'S STACKS OR ARCHES: The arched rocks visible at the end of the cape of Cabo San Lucas are locally known as "The Friars". Geologically, the Friars are stacks—isolated arch-like rocky islands detached from the tip of the peninsula by wave erosion. At some time in the future "The Friars" will be completely destroyed by wave erosion along joints. Waves pounding against a wave-cut cliff produce various features as a result of the differential erosion of the weaker sections of rock. Wave action may hollow out cavities or sea caves in the cliff, and if this erosion should cut through a headland, a sea arch is formed. The collapse of a roof of a sea arch leaves an isolated mass of rock called a stack. The arches are similar to those seen at Big Sur in northern California. This area is commonly referred to as "Land's End".

5 The fairly prominent terrace on the high hill is also part of the main point. It reflects a higher stand of sea level at 10 meters. There is a large coastal dune field at about the 10 meter terrace level which extends up the coast for approximately five kilometers.

3 The highway drops down close to sea level.

2 The highway crosses a wash. Take the road to the right to Todos Santos. To continue all the way to the cape follow the divided highway around the harbor.

In the flats which surround Cabo San Lucas, Screwbean Mesquite is the dominant plant accompanied by scattered Lomboy, Cardon, both Pitaya Dulce and Agria, Cholla, and Tamarisk.

The granitic rocks between Hotels Solomar and Finisterra are highly jointed and contain large **xenoliths**.

SURROUNDING ROCK IS METAMORPHOSED

MILES BELOW THE SURFACE

MAGMA = MOLTEN ROCK

PIECES OF SURROUNDING ROCK FALL INTO THE MAGMA AND PARTIALLY MELT FORMING ZENOLITHS

XENOLITHS: During the intrusion of magma, the molten rock melts and pushes its way into the surrounding rocks. Pieces of the surrounding rock are broken off and sink into the magma. Some of these pieces are partially melted and the resulting piece is generally richer in iron and thus darker than the igneous rocks of magma. These pieces are called xenoliths. Miles of uplift and erosion raises everything to the surface.

XENOLITH IN GRANITICS AT CABO SAN LUCAS

JOINTS AND XENOLITHS IN GRANITICS AT CABO SAN LUCAS

If you take a short hike along the beach and climb over the low headland, you will be at land's end and able to cross the Peninsula on a beach.

PACIFIC BEACH AT LANDS END

GULF BEACH AT LANDS END

TOTAL ECLIPSE OF THE SUN IN SOUTHERN BAJA
JULY 11, 1991

PACIFIC BEACH AT LANDS END

COMORANTS ON ROCKS

Log 10 - Cabo San Lucas to La Paz via Todos Santos [157 kms = 97 miles]

From Cabo San Lucas the Highway heads northward traveling largely on the dissected surfaces of alluvial fans from the granitic Sierra la Laguna. Extensive dune fields and supra tidal flats are near the road where it lies close to the ocean. Near Todos Santos the road climbs through several low passes in the granitic and metamorphic rocks. At Todos Santos the road turns inland to travel on a dissected alluvial surface along a major fault close to the granitic hills of the Sierra la Laguna.

123 There are two routes from Cabo San Lucas along Highway 19 to Todos Santos and La Paz. Retrace the route back to the Pemex station and turn left on Highway 19 north or take the business route from town which is Paseo Morelos.

From Cabo San Lucas the highway goes northwest on alluvial fans toward the granitic massif of Sierra la Laguna. In this area the highway passes through the Cape phytogeographic region. The vegetation along the highway consists of a heavy thorn forest underbrush dominated by species of Acacia, Organ Pipe Cactus, Cardons, Elephant Tree, Yucca Vine, Lomboy, Palo Mauto, and several species of Cholla.

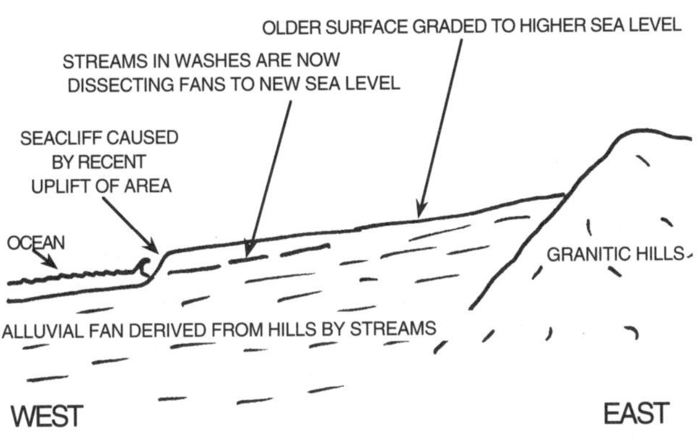

115 The highway undulates through an area of mature topography cut into the alluvial slopes. The lack of outcrops in this area is the result of the deep tropical weathering of the granitic rocks and subsequent burial of the outcrops by alluvium. The present streams are cutting into the fans and will eventually establish another surface a few 10's of feet lower than the present surface.

113 Cardons have become the predominant taller vegetation with a sparse desert "understory" of Leather Plant.

CARDONAL

The highway passes through several roadcuts in the highly weathered granitic rocks. Somewhat fresher granitic rocks are visible in the washes of this region.

111 The highway crests a pass with sweeping views of the Pacific Ocean and descends a long grade on the alluvial fan surface toward the pacific coast.

107.1 As the highway crosses a wash, excellent exposures of granodiorite can be seen in the lower part of the wash.

107 The highway climbs a long grade over the surface of a dissected alluvial fan.

104.5 As the highway crests over another small pass, the Pacific Ocean and many active coastal sand dunes are visible to the west.

103 The species composition of the vegetation is essentially the same as that listed at Kilometer 123, but the trees are noticeably shorter, rarely exceeding 3-4 meters in height. Lomboy and Leatherplant dominate the understory of a short "impoverished forest" of Elephant Trees, Cardon, and a few scattered *Acacias*.

Over 8,000 species of euphorbs grow around the world. Some furnish food and valuable oils, while others are of ornamental or medicinal value. The two species of *Jatropha* commonly seen along

Baja's highways on plains, hillsides, mesas, and sierras are *J. cinerea* and *J. vernicosa*. The sap of *J. cinerea* is highly astringent and is said to stanch bleeding wounds, prevent chapped lips, and, unfortunately, permanently stain clothing. Specimens of Lomboy grow from Punta Prieta on the Pacific side of the peninsula and Bahia de

LOMBOY

STANCH IT WITH LOMBOY OR NATURE'S "CHAP STICK": The low shrubby Lomboy (*Jatropha sp.*) seen along the highway in this part of the Magdalena Plain is a member of the spurge family. Members of this family usually produce a milky acrid sap.

Leatherplant Another euphorb, *Jatropha aurea*, a low shrub, grows in this area.

Elephant Trees of this area are represented by the unrelated species of the two genera *Pachycormus* and *Bursera*. Because they store water in the cortical cells of their elephantine trunks, Elephant trees grow luxuriously despite the extreme aridity of this region. The *Bursera* are easily distinguished by the incense-like odor produced by crushed leaves. *Pachycormus* tissues are odorless.

Palo Verde is a green trunked tree which produces legume-like "bean" pods. However, Palo Verde is in the Senna family, not the legume family. Like Lomboy, this tree is also normally leafless, an adaptation which reduces water loss by transpiration (a process of water evaporation through thousands of leaf pores) during dry periods. The green trunk of the Palo Verde is a cladode; that is, it's a photosynthetic stem which acts like a leaf and performs photosynthesis in the absence of leaves.

Palo Adan is a relative of the Ocotillo which is commonly seen in the northern deserts of Baja. However, Palo Adan has thicker branches, a trunk, smaller flowers, and normally grows from Parallel 28° south to the Cape Region on the clay and granitic soils of alluvial plains.

PALO ADAN

Ramalina is an epiphytic foliose lichen formed by the combination of algal cells which live inside the tissues and cells of a fungus in a mutualistic symbiosis. The algal cells provide sugar to the host fungus which utilizes it as a source of energy. In return the fungus provides protection and the water and carbon dioxide utilized by the algal symbiont during photosynthesis. Other species of lichen form living crusts on rocks in this area. Crustose lichen look like variously colored "splashes" of paint. A lichen is a symbiotic relationship between an algae and non-photosynthetic fungus. Due to a very specialized life system, they can exist on bare rock and obtain food from the air, sunlight, rain water, and rocks. They are instrumental in the breakdown of rocks by their production of weak organic acids which result in the production of soils. The lichen also keep the rock moister and promote more local chemical weathering. Lichens are common desert dwellers and are often the dominant vegetation in certain desert ecosystems.

98 As the highway crosses a low wash vegetated with Broom Baccharis, it passes roadcuts of mixed granitic rocks with pods of metamorphic rock cut by dikes.

99 The mountains in view along the highway to the right are the granitic Sierra la Victoria.

98 As the highway crests a low pass, there is another view of the Pacific Ocean. Numerous active and stabilized dunes near the coast are seen to the west.

The vegetation in this region has become sparse. The taller plants are Elephant Trees and Cardon, shrubby, leafless Lomboy and Leather Plant dominate the "understory".

96 For the next few kilometers the highway parallels the coastline which ranges from 1 to 2 kilometers to the west of the highway. Along this stretch there are almost continual views of the Pacific Ocean and its fringe of active, and stabilized sand dunes.

91 The highway drops into another wash where the dirt road to the west leads approximately one kilometer down the wash toward the Pacific and ends at a beautiful sandy beach with rocky points and headlands composed of granitic rocks.

82 From this point an alluviated surface stretches to the beach. It is truncated at the 10 meter terrace level above sandy beaches. A beautiful

beach is located at the bottom of this gentle grade near Las Cabrillas. However, it is unsafe to swim here because the surf in this area is dangerous.

Granitic rocks and mixed metamorphic rocks are occasional seen in the roadcuts. The highway in this area was built on a base of very coarse decomposed granitic soil that developed on the alluvial fans which stretch west to the Pacific. The granitic sands originated in the high Sierra la Victoria and were brought down by stream action and runoff.

79 View of the ocean and several ranchos which are surrounded by Fan Palms and fruit trees.

75.5 A storm beach berm has trapped water in the low land behind it forming a supertidal flat.

A tall Cardon and a few scattered Mangrove Trees are growing behind the berm. Cardon are also growing on the surrounding well-drained drier hillside slopes. Due to the scarcity of water, the vegetation of the slopes is shorter and not much more than one meter high.

74.5 A good short road approximately 100 meters long leads to the west to the beach. A short walk in this area makes possible a close inspection of the coastal storm beach berm, supertidal flats, and alluvial surface.

73.7 The highway crests a low pass through gneiss. The view to the north is of several kilometers of the Pacific coastline beaches.

71 Ball Moss grows profusely on the Palo Adan.

69.2 Another good view of a supertidal flat (Salitrales) is visible. Salitrales are often dry a good part of the year. Further north on the west coast of Baja they form hard surfaces that are excellent to travel upon. However, when they flood during the summer rainy season or very high tides, they become slick and "soupy" and very unusable.

68 The highway continues to traverse the alluviated fan surface with numerous views of Punta Lobos. Gneisses and other metamorphic rocks form Punta Lobos.

52.2 Todos Santos is a relatively small agricultural village developed in a lush green valley near the Pacific and is located close to the Tropic of Cancer. The town began as a farming community and a mission visiting station in the early 1700's and finally gained mission status when the Mision Todos Santos de Santa Rosa was built in 1732. The tall shiny-leafed trees which line the highway into Todos Santos are Mango Trees. At the sugar cane mills of Todos Santos, cane is delivered and crushed which produces a sweet liquid. The liquid is cooked into a dark liquid which is poured into hollowed-out cone-shaped forms which produce a coarse grade of raw dark-brown sugar called "panocha".

50.8 An anticline is visible in the metamorphic rocks in the roadcut on the right.

To the north away from Todos Santos, the highway roughly parallels the Sierra la Victoria Fault Zone that continues across Baja Peninsula and passes just east of La Paz. To the left is the faulted graben of Llanos de Santo Tomas. The Todos Santos Graben is underlain successively by Pliocene sedimentary rocks, Miocene volcanics and marine sedimentary rocks, and ultimately by a Cretaceous syncline which is truncated by the Sierra la Victoria Fault.

45 As the highway rounds a curve, the foothills of the northern Sierra la Victoria can be seen. The frontal fault of the Sierra la Victoria cuts diagonally toward the highway from right to left.

39 The highway descends into a large arroyo through exposes of granitic rocks. A thin terrace covers the granitic rocks while coarse blocks of granitic debris are visible downstream.

Locally the massif of the Sierra la Victoria has been planed off into a pedimented surface. The pediment is seen as a gently sloping erosion surface developed by running water at the base of the abrupt and receding front of the Sierra la Victoria. The pediment is underlain by a bedrock of granitic and metamorphic rocks mantled with a thin discontinuous veneer of alluvium derived from the upland massif of the Sierra la Victoria. This thin veneer passes over the fault and becomes quite thick on the down-dropped side of the fault. There are a complex series of faults in this frontal fault zone.

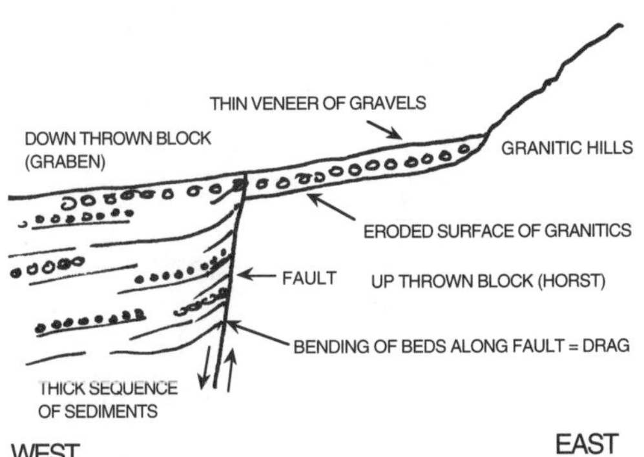

29 There is a good view to the right rear of the massif of the Sierra la Victoria. The highway continues to the northeast across the flat alluviated surface for a number of kilometers and then turns to the east toward La Paz. As the highway traverses this flat featureless plain, it approaches the smaller massif of the Sierra las Calabazas. This small range, which resembles bald heads, is visible to the right of the junction of Highway 19 and Highway 1.

34 Watch carefully as you drive this section of the *Baja Highway,* and you may see a Cuckoo Bird (*Geococcyx californianus*) run across the road.

The **Greater Roadrunner or Correcamino Californiano** is the large terrestrial Cuckoo Bird which may be seen running through the chaparral, desert scrub, and thorn scrub throughout the entire peninsula. The most identifiable field characteristics of the roadrunner are its bushy crest, brown and white streaked back, and long black, white-tipped tail. A close look at the eye will reveal that it's flanked by beautiful red, white, and blue feathers. The food of this fleet-footed predator consists of anything that moves (primarily lizards, snakes, and insects). This birds scientific name translates to "California earth tail" referring to its running habit and long tail.

0 Junction. Highway 19 north leads to La Paz; Highway 1 south leads to Cabo San Lucas. From the junction it is approximately 34 kms. to La Paz.

CASTING NET FOR BAIT

JASON TAKING COVER PHOTO

Log 11 - Laguna Chapala to San Felipe via Gonzaga. [215 kms = 133 miles]

NOTE: The kilometer markings may change. Use the main points as relative references.

0 The highway proceeds east, across the north end of Laguna Seca Chapala, on a roadbed which is elevated above the dusty lake bed.

2.6 This is the Old Rancho Laguna Seca Chapala which is now abandoned. The site was chosen because of the mines, which are low on the hill of granitic rock to the north, and because the old main road down the peninsula intersected another crude road going east to the mines at Las Arrastras and to Bahia San Luis Gonzaga. For the next 5 Kilometers the highway will travel through a granitic and metamorphic terrain.

7.7 After crossing a small fault, the highway climbs out of the arroyo over a hill in a belt of metavolcanic rocks which have been faulted between granite, slate, and schist.

8.6 The highway crosses first into granitic rocks, then into metamorphic rocks, then makes a bend up a canyon in the metamorphic rocks.

The vegetation consists of Creosote, Brittle-Bush, Garambullo, Cardon (small and sparse), Cirio, Deadly Nightshade (opportunistic along the road), Ocotillo, and shrubs such as Rabbit Brush and Purple Sage (as in Zane Grey's "Rider's of the Purple Sage").

9.7 Cross over a little divide in the metavolcanic rocks, and begin to descend a canyon in metavolcanic rocks dipping to the northeast at about a 70 degree angle.

10 The hill to the right is cut by several small quartz dikes. The quartz is weathering out and scattering down the sides of the hill.

11 Crest of a grade.

13.6 The highway follows down a fairly broad and rather steep canyon into progressively older parts of the metavolcanic section.

There are bigger Ocotillo, Cardon, and Elephant Trees with Teddy Bear Cholla in the bottom of the canyon. On the sides, the Elephant Trees become sparse, and the Cardons drop out leaving Ocotillo, Old Man Cactus, Buckwheat, Barrel Cactus and Agave.

The following bird species are typical of the San Felipe Desert:

BIRD NAME	LIKELY LOCATION
Abert's Towhee	Desert woodlands and streamside thickets
American Kestrel	wires and fence posts
Amer. White Pelican	Gliding along the shore
Anna's Hummingbird	red or yellow tubular flowers
Bendire's Thrasher	Flies from bush to bush, feeds on ground
Cactus Wren	On cacti
Calif. Brown Pelican	Gliding along the shore
California Quail	On the ground
California Thrasher	On the ground
Costa's Hummingbird	feeding on red flowers
Crissal Thrasher	Secretive, hiding in brush
Gambel's Quail	On the ground in desert scrublands and thickets
Gila Woodpecker	Nests in holes in giant cacti
Greater Roadrunner	Crossing the highway
Ladder-back Woodpecker	Within the desert, sometimes nest among the agave stalk
LeConte's Thrasher	In sparse vegetation, flies when necessary
Loggerhead Shrike	wires and fence posts
Red-Tailed Hawk	Tops of telephone poles and fence posts
Scrub Jay	In chaparral and woodlands
Turkey Vultures	Soaring in the skies or feeding on carrion
Vermilion Flycatcher	Streamside shrubs and wooded ponds
Western Meadowlark	Fence posts and fence wires

14.5 As the highway exits the canyon, it crosses a high angle normal fault onto uplifted alluvial fan material.

15.4 The highway follows along the faulted edge of a tonalite body with an alluviated valley to the right.

16.6 For the next 5 Km the highway passes through roadcuts and crests a small pass in tonalite and metamorphic rocks.

17 Enter a wide ampitheater of Creosote Bush, Ocotillo, Old Man Cactus, and Palo Verde with Smoke Trees in the washes.

20.5 The tonalite ahead and to the left is capped by rhyolite. Most of the terrain in this area and the near distance is tonalite. In the far distance the hills are tonalite, slates, and metamorphic rocks interspersed with thin basins of alluvium. The vista opens up to the north. While the Gulf cannot be seen, the view is in the direction of Bahia San Luis Gonzaga.

21.3 The kilometer markings start at "0" again at this junction. The main highway continues to the left. The road to the right leads to Puerto Calamuje.

The main highway crosses an open alluviated plain with tonalite bodies visible on all sides. The low hills, about 0.5 Km, to the left represent an eroded fault scarp formed by a fault paralleling the highway. The tonalite exhibits some of the typical spheroidal weathering often characteristic of desert terrains. (See 3:153.5)

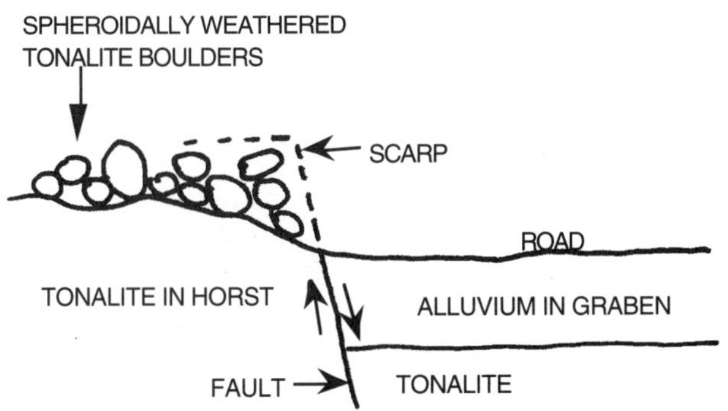

The vegetation of the large amphitheater between Km 0 and Km 5 is Creosote Bush. Growing on the large boulder hillsides is Creosote Bush and occasional Cardon, Ocotillo, Dodder on Elephant trees, and Brittle Bush.

4 To the north is a view of the rhyolitic rocks covering the Sierra's north of Bahia San Luis Gonzaga. To the left there is an outcrop of hyolite covering basalt on top of the hills

6 The highway bends to the left to the Old Rancho Las Arrastras. Just past the ranch the highway drops into a rocky granitic arroyo. Directly ahead are basalts unconformably overlying tonalite.

7 From Rancho Las Arrastras, the highway now follows down a branch of the Arroyo las Arrastras in the tonalite. The small reddish-brown mammiform conical hill, to the right, is covered with rhyolite and surrounded by a talus slope.

8 The darkly banded hills to the left, with steeply dipping layers, are metavolcanic rocks.

11 The rocks of this area are predominated by jointed tonalite. Xenoliths, the darker, partially melted remnants of the rocks which originally enclosed the tonalites, are visible in the tonalite. The tonalite often weathers into hollow cavernous boulders. Good examples of cavernous weathering are exposed here and at Km 12.

CAVERNOUS WEATHERING: For any one of a number of reasons part of the tonalite boulder is more protected and does not dry as fast. As a result, chemical weathering proceeds at a faster rate resulting in the alteration of feldspars to clays. This change pries the grains from the protected surface and produces an even more protected area. The wind blows the loose grains away gradually forming the caves.

12.5 The highway drops down into the main part of the Arroyo las Arrastras

This is an excellent spot for photographing an unconformity of tonalites covered by rhyolite.

14 As the highway crosses and recrosses the Arroyo las Arrastras it begins to follow the surface of a gravelly Quaternary terrace.

15.5 The highway drops into another wash vegetated with Cats Claw Acacia and Cheese Bush. The Wash Woodland vegetation is similar to that of the Anza-Borrego Desert in southeastern California. The predominant floral cover consists of Creosote Bush, Atraplex, Burroweed, Garambullo, Cardon, Ocotillo, and Mistletoe laden Palo Verde (*See* 16:15.5).

The highway again climbs up onto the high, gravelly Quaternary terrace with beautiful stands of Ocotillo and tall Elephant Trees with sand Verbena growing along the roadside.

The area to the left is Miocene andesite that has been faulted into a series of benches resembling a series of giant steps leading down toward the Arroyo las Arrastras.

18 The Quaternary fluvial terrace material that the road is following parallels the present-day San Francisquito Creek. It probably represents part of the ancestral bed of the creek when sea level was higher. When sea level lowered the area was abandoned by the creek leaving a relatively smooth terrace. The creek is vegetated by a vast expanse of Ocotillo, Brittle-Bush, Burroweed, Creosote Bush, and scattered Teddy Bear Cholla.

21.1 This road is the turn off to Punta Final.

23.5 This is the first view of Ensenada San Francisquito and Bahia San Luis Gonzaga. The far point directly ahead which looks like a low hill is Isla San Luis Gonzaga. The small point to the left is Punta Willard the north point of Bahia San Luis Gonzaga. The island and Punta Arena separate the main part of Bahia San Luis Gonzaga from the Ensenada San Francisquito.

27 The vista ahead reveals good views of Isla San Luis Gonzaga, Alfonsinas airstrip on Punta Arena, Bahia San Luis Gonzaga, and Ensenada San Francisquito. In the far distance, to the northwest, the gray point is Isla San Luis.

27.5 The highway passes through a wide part of the Arroyo las Arrastras. This crossing provides evidence that the entire area was uplifted. The stream has incised about 10 meters below the level of the braided stream drainage of the old alluvial fan. The upper surface probably represents the surface graded to the Sagamonian 5E high stand of sea level about 125,000 years ago. This high stand has been documented with coral dates at Mulege (Ashby and Minch, 1989).

35 This turnoff leads to Alfonsina's airstrip.

36.5 The highway passes through a roadcut blasted into andesite material. The ground here is littered with dark colored cinders.

From Bahia San Luis Gonzaga north to San Felipe the highway passes through the San Felipe Desert subdivision of the Desert Phytogeographic Region (*See* 14:108).

37.2 The kilometer markers descend to San Felipe. Km 37.2 is 154.3 Kms south of San Felipe.

The layered nature of the andesitic volcanics is very obvious on the north point of San Luis Gonzaga. The hills to the north are andesitic volcanics. Just around the edge of Punta Bufeo is the gray mass of Isla San Luis. Ash-gray cinders litter the ground.

147.5 The highway rounds a corner and drops slightly into a flat alluviated area. The Pliocene marine sedimentary rocks seen here form a "badlands-like" topography under the alluvium. The highway is flanked on both sides by andesite hills.

146.7 The mainland range to the left is all rhyolite. This roadcut is in the altered sedimentary rocks of the Pliocene Salada Formation.

At the crest the vista opens with the gulf and a number of volcanic islands in the distance. The islands, in succession, are: Chollito, which looks like a divided island; Isla Lobos, the pointed one; and, to the left, El Muerto, which almost looks connected to the mainland. Isla El Huerfanito ("the Little Orphan") is not yet in view.

On the alluviated plain the reddish and yellow beds of the Pliocene marine sedimentary rocks of the Salada Formation are exposed to the left in the gully below. They are overlain by terrace material. The bouldery outcrops on the higher hills behind the Pliocene are spheroidally weathered tonalite. The andesite volcanics drape over the older formations. The highest hill is a small basalt shield cone called Cerro el Portrero.

The highway descends through the Pliocene sedimentary rocks of the Salada Formation onto the flats, with very low hills, underlain by Pliocene rocks, in the near distance ahead.

The vegetation covering the Pliocene marine terrace consists of Ocotillo, Palo Adan, and Creosote. It is sparser here than it was at the vista point back at the crest of the hill (Km 146.7). Smaller Smoke Trees, Rabbitbrush, and Burroweed predominate in the washes along this stretch of the highway.

144.9 Pliocene beds are exposed in the roadcut at the edge of this wash. The low hills to the left exhibit exposures of the Pliocene Salada Formation. Volcanics are visible to the right.

Along this stretch, the highway alternately climbs on the low fluvial terrace, drops down into a wash, and returns to the terrace.

The vegetation has dramatically changed. Except for the Creosote Bush and several species of annual grasses, almost all of the plants mentioned at Km 146.7 have disappeared. A few very small Ocotillo can be seen growing on the rises between the washes.

141.7 The north end of Isla San Luis is directly east of the highway

ISLA SAN LUIS consists of several low rings of pumice (the gray material) with a very obvious obsidian dome (the darker material). The same geological conditions exist at Mono Craters in central California. The south end of Isla San Luis is a Maar, an explosion pit consisting of a large pumice ring with no dome. The southern point of the island is the west rim of the explosion pit. The islands cliffs were formed by the sea eroding the soft pumice. The obsidian on the larger dome was dated using obsidian hydration rates and found to be as young as 100 years. The darker material at the north end of the island is another obsidian dome.

141.2 Turnoff at Punta Bufeo. At the north end of Punta Bufeo, the vegetation consists almost entirely of Creosote Bush, a few Brittle-Bush, annual grasses, and Ocotillo which appear mostly on the edges of the washes or occasionally scattered about on the Fan surface.

138.6 The excellent view northward across the alluviated fan is of Isla San Luis and most of the other small offshore islands located to the north of Punta Bufeo.

The tonalite hills ahead are covered with the brown mottling effects of desert varnish. The light colored material cutting across the hills in streaks are dikes.

134 The highway crosses the Arroyo Mal de Orin. The prominent wash woodland vegetation of this arroyo consists of Smoke Tree, Elephant Tree, Horehound, Princess Plume, Spanish Bayonette, Yucca, Pencil Cholla, and mistletoe laden Palo Verde (*See* 16:16.5).

132.8 Garambullo and Cheesebush are the tall floral dominants. An interesting member of the squash family, a bush with a yellow-orange, trumpetlike flower called a Devil's Claw or Unicorn Plant can occasionally be sighted from the highway.

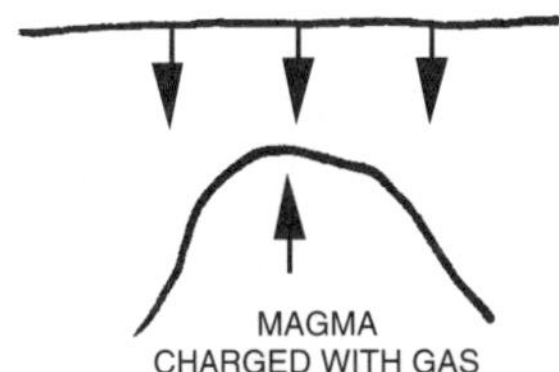

1 AS THE HOT MAGMA PUSHES TOWARD THE SURFACE THE WEIGHT OF THE OVERLYING ROCKS PREVENTS AN ERUPTION

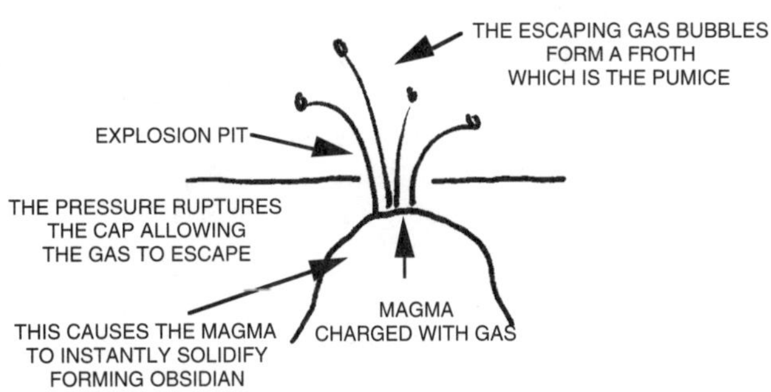

2 THE ERUPTION OCCURS WHEN THE MAGMA PRESSURE EXCEEDS THE WEIGHT OF THE OVERLYING ROCKS

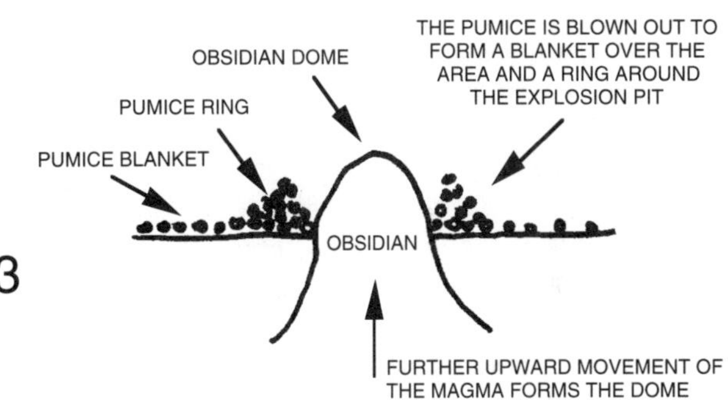

3

133 The highway has been traversing an active alluvial fan for several kilometers. The low ridge ahead (15 to 20 meters high) parallels the surface of the fan. This ridge is an older terrace deposited when sea level was higher.

131.1 The vegetation has changed dramatically. The ground is almost entirely obscured by a thick cover of Brittle-Bush. In the springtime the area is illuminated by the bright yellow flowers of the Brittle-Bush. There are a few scattered Creosote whose numbers increase toward the beach. The highway approaches the end of the coastal alluviated Pliocene marine terrace. The low hills ahead are volcanic. The tonalite on the left is cut by numerous nearly vertical, light-colored, ribbon like dikes. A foothill "screen" of dark metamorphic rocks is exposed close to the volcanic hills forming a low belt of darker colored, steeper hills than the volcanic hills ahead.

127.6 The volcanic island of Isla el Muerto lies directly offshore. The white is bird guano.

126.4 The Brittle-Bush, seemingly the only plant growing here, looks like little gray "buttons" all over the hills. Because each plant has its own nutrient and water requirements, they space themselves naturally, appearing to have been planted in rows. At lower elevations, between the highway and the gulf, Creosote and Ocotillo are occasionally growing among the Brittle-Bush.

124.8 The highway crosses a rather narrow but deep wash (Arroyo Miramar). Hard, well cemented, fanglomerates are exposed in this arroyo.

124.5 The view to the south is of Isla el Muerto with sandy gulf coast beaches stretching south-eastward. On a clear day one can see all the way to the south to Punta Bufeo.

Magnificent Frigate birds are commonly seen soaring off shore (7:94.3).

122.3 After passing a very small cove the highway rounds a corner to skirt a large cove with a beautiful curved, cobble, shingle beach with a number of storm berms, and the ruins of a small settlement behind them. The cove is a beautiful place to camp. Shorelines in volcanic terrains are usually covered by rounded pebbles and cobbles. These cobbles are often flattened and stacked like shingles on the beach face.

116.3 A sudden change has occurred in the vegetation. There has been an increase in the densities of Cardon, Ocotillo, Garambullo, Palo Verde, Elephant Trees, Cheesebush, Ocotillo, large Acacias, and a variety of low annual grasses. Smoke trees are generally found in washes, because their seeds have a very resistant seed coat which must be scarified by sand grains as water flows down the washes (*See* 13:43.5).

109.9 The view ahead is of the village of El Huerfanito and the white, bird guano covered Isla El Huerfanito. Just to the left of El Huerfanito the view is of the highway climbing up the first of the steep grades of the Cuesta La Virgin. The dark colored hill midway up the coast, behind the grade, is the basalt cone, Volcan Prieto.

107 Climbing the grade provides a view of a series of stripped dip-slopes, dissected by streams, reaching to the ocean. Below the slopes the streams valleys form steep sided fjord like embayments .

The Brittle-Bush again resembles little gray "buttons" on the dark volcanic material. Atriplex and Wild Buckwheat are seen infrequently scattered among the Brittle-Bush.

ON SECOND SPARE

106.2 The highway crests the first little grade then drops into Arroyo El Huerfanito, for a kilometer or so, before climbing the main part of the first big grade.

104 At the top of the roadcut near the crest of the first big grade is a pinkish rhyolite with a 15 to 30 cm. baked zone. Below that is a very muddy rhyolitic or andesitic lahar, which looks like a mud flow, and a layer of basaltic cinders.

The highway eventually drops down into a major wash (Arroyo Heme) the first north of the first grade. It should be possible to go down the wash to the beach. The highway climbs up out of the wash to a view of a nice rocky beach with a storm berm. As you drive along the ocean, watch for shrimpers out of mainland Mexico. The boats usually have a white hull and long seine net booms extending outward over both sides. The net booms and their seine nets look like huge grasshopper legs from the side and like wings from the front.

VIEW TO SOUTH OF COAST AND ISLANDS

102.1 The small dark hill is a Quaternary basalt cone.

103.8 Near the top of the grade, where the highway crosses to the east side of the ridge the view to the north and west at the volcanic tableland looks deceptively flat, because you are looking at the dip-slope. Where this flat face is cut by canyons it is revealed as being a steep slope.

101 Halfway through the east side of the ridge is a road cut with a rhyolite flow overlying a very thin mudflow. There is a slight angular unconformity where some deformation and erosion took place between deposition of the beds. Below that is a 0.7 m. thick cinder blanket of basaltic fragments.

After crossing over the edge there is a turnout at a wide spot with a panoramic view to the south. In the far distance is Isla San Luis Gonzaga, the edge of Punta Final at the south end of Bahia San Luis Gonzaga, the highway where it seems to disappear at the north point of Bahia San Luis Gonzaga, Isla San Luis, Ensenada San Francisquito, and the high parts of the Sierra La Asamblea south along the main peninsular divide. To the right (west) of the view is the stepping-up of the dip-slopes of the rhyolitic rocks with a flat slope, a steep rise, flat slope, steep rise, etc., showing where each successive layer of the rhyolites has been stripped off.

97 The second grade descends until it nears sea level on the alluvial surface. After passing the rocky beach the highway climbs another hill.

95 A stop at the top of this hill will provide another spectacular panoramic view to the south. All of the near coast gulf islands are clearly discernible. Along the right side of the highway, at the very southern end of the large basaltic cone, a large sandy tidal flat, often below sea level, is seen. Fresh water from the arroyo or sea water can fill this entire area during high tides and storms.

94.7 The highway crests here at about 140 meters above sea level then descends and begins undulating again following a terrace level. The little point in the distance is the north point of Puertocitos.

89 The hills nearer the highway are covered with white *Eriaganum* - another species of wild Buckwheat. The other hills are covered with Brittle-Bush, Ocotillo, and Mistletoe laden, trees.

There is a road going down to a small settlement on a fine gravelly beach. A tombolo extends to a small, near shore, rocky island at low tide. A dip-slope has created a barrier with many excellent, beautiful tide pools.

85 The grade just south of Puertocitos provides a nice view of the harbor. Look for shrimp boats, Magnificent Frigates, Brown Pelicans, and Gulls in the bay.

80 **PUERTOCITOS** is an elongated, northwest-southeast oriented, linear bay formed by faults along both sides of the bay. The rocky shores (horsts) on the west and east are formed by uplift along two faults with the down-dropped (graben) block forming the sandy beach in the middle. Because of a 8 meter tidal range in this area, the high tide on this beach will come up to the foundations of the buildings. At low tide the bay is almost devoid of water. Toward the edge of the point, about 3/4 of the way down the east side of the bay, there is a hot spring located along the fault. At low tide the steam can sometimes be seen rising from the area.

The main highway closely follows the fault line on the west side of the graben. The cliff on the left side is a fault scarp. More of the fossiliferous Pliocene beds are exposed near the airport. The highway climbs out of the graben through a low pass in the hills. Most of the dark rock on the beach side of the highway is rhyolite.

PUERTOCITOS AT HIGH TIDE

PUERTOCITOS AT LOW TIDE

79.5 Beautiful rocky points and pocket coves are present along the highway where the resistant rhyolite beds dip into the ocean.

75 The highway crests a low pass in the hills and drops down through volcanic rocks. Several levels of alluvial fans are now visible, each of which were graded to sea level at their toes at one time and which now, due to the uplifting of the land surface, end in sea cliffs. The relative ages of some of these have been estimated by calculating the rate of the drop in sea level.

71 The mustard yellow hills on the left are Pliocene marine beds capped by an alluvial fan.

68.4 The highway climbs up through more yellow beds and turns through a low roadcut.

The vegetational cover here consists of Ocotillo, Elephant Trees, Creosote Bush, Mesquite with Mistletoe, and Cheese Bush.

BURROS - BURSAGE - RHYOLITE

68 North of Puertocitos the highway drops into the large Arroyo Matomi and runs "as straight as an arrow" across alluvial fan material.

Most of the hills in the distance are volcanic capped. The tonalite plutons of this area are covered by fluvial and marine Miocene and Pliocene volcanic and sedimentary rocks. The large hill on the horizon at about 10 o'clock is the high point of the Sierra San Fermin, a volcanic eruptive center composed of rhyolite. The high granitic crest of the San Pedro Martir and Picacho Del Diablo can be seen in the far distance at about 11 o'clock.

61 The low hill on the horizon to the southwest (left rear) is Picacho Canelo, a rhyolitic volcanic eruptive center.

58 Toward the end of the plain, the highway climbs on the terrace in the foothills of the Sierra San Fermin, makes a deviation to the left, and crosses more washes and terraces.

54.2 The side road to the right leads down to Playa Cristina where a small lagoon and a coastal dune field have developed.

53 The highway drops from the elevated terrace back onto the alluvial fan and crosses Arroyo de Chale.

48.8 Campo Coloradito. The mountain range to the left is the Sierra San Felipe. This range continues northward to San Felipe.

37 The vegetation changes to a Creosote Bush/ Ocotillo community typical of the San Felipe Desert (*See* 14:0 and 14:108).

30.6 Bahia Santa Maria. The road to the left goes to Agua de Chale and the sulphur mines. The sulphur was brought to the surface by vapor and hot water along a fault line. The hot spring or fumarole activity has altered the rhyolite beds and deposited the sulfur as small yellow crystals and crusts in the cracks and cavities in the rocks.

30 The coastline here has a wide mud-flat in back of the beach behind the storm berm. The high hill to the north, between Punta Estrella and Punta Digs, is Cerro Punta Estrella, a largely tonalite hill with some prebatholithic metamorphic rocks along its flanks.

The vegetation stabilizing the coastal dunes is primarily Creosote Bush, Mesquite, Atriplex, Smoke Trees (in the washes), Garambullo, Palo Adan, Brittle Bush, and *Acacia*.

20.5 Road to supertidal flat of Laguna Percubu.

18 The tonalite pluton of Cerro Punta Estrella straight ahead is becoming impressive. It looks as though the highway is heading straight into the mountain range. In this area everything seems to be at a great distance.

17 There is a cove like indentation in the tonalite hills which contains dark metamorphic rocks cut by numerous light colored dikes.

A dune field is on the north side of the large tonalite hill. The highway passes through the hummocky topography of stabilized dunes which have been blown from the coast by the wind.

6.8 At Hotel Farro Beach the highway drops behind and follows along the edge of the semi-stabilized dune (10 to 15 meters high).

The Sierra San Pedro Martir and Picacho del Diablo, the highest peak in Baja California (3,115 meters, 10,126 feet) are in view on the left front.

The dunes in this region are moving. Some of the vegetation growing on or near these dunes consists of Ephedra, Creosote Bush, Ocotillo, Atriplex and Cat's Claw Acacia. These last two plants are deep-rooted and the dunes hold moisture so they can survive in this area. On hummocks the sand is held in place by the roots. As the sand builds up the tree grows higher.

0 The highway divides. The road to the left goes to the airport; the one to the right climbs on the dunes and goes about 5 Kms. into San Felipe.

As the highway crests over a dune the beautiful expanse of Bahia de San Felipe and the town of San Felipe comes into view. The coastline here tends to be muddy because of the Colorado River and the volcanic terrain which produces finer grained material. As you drive down the highway, along the crest of this stabilized dune, you can see the sand being blown across the highway because there is little brush to stop it.

The highway soon crosses an arroyo. The right point of the harbor is Punta El Machorro consisting of tonalite. The slightly higher hill to the left is Cerro El Machorro, consisting of granodiorite. The low-lying hills in the foreground, behind the city, are prebatholithic carbonaceous rocks of undetermined age.

Where to from here: Follow the reverse of Log 14 to Mexicali or at Km 140 turn west and follow Log 12 to Ensenada.

REMOTE AREA OF BASINS AND RANGES

Log 12 - San Felipe to Ensenada via Valle Trinidad [245 kms = 152 miles]

San Felipe to Baja Highway 3 *- The Highway climbs onto uplifted alluvial fans and passes through a series of steep granitic and metamorphic hills. For several kilometers north of the hills the Highway undulates across the dissected uplifted fans. The granitic Sierra San Pedro Martir form the high mountains on the far left. The largely granitic Sierra San Felipe form the nearer desert varnished foothills. As the Highway crosses the alluvial fans, the last views of the Gulf of California fade into the distance across the supertidal flats of the Salinas de Omotepec. The Highway crosses several of these dune areas with sand blown from the Gulf and supertidal flats. The Highway continues for many kilometers along the alluvial fans descending from the rugged Sierra de San Felipe with views ahead of the rugged Sierra Pintas.*

Baja Highway 3 to Valle Trinidad *- From the intersection of Highway 3 the Highway heads west across the uplifted dissected alluvial fans, past the rugged hills of granitic and metamorphic rocks of the Cerro El Borrego and the Sierra de San Felipe, towards the crest of the granitic and metamorphic Sierra San Pedro Martir. After skirting the north end of the alluviated Valle San Felipe graben the Highway enters a rugged, steep area of granitic and metamorphic rocks in San Matias Pass at the end of the Agua Blanca Fault Zone. The Highway then follows the trace of the Agua Blanca Fault Zone at grade through the steep hills of granitic and metamorphic rocks to the broad alluviated area of Valle Trinidad. Miocene volcanic rocks cap the mesas to the north of the pass.*

Valle Trinidad to San Salvador *- At Valle Trinidad the Highway turns and climbs a steep grade on a scarp of the Agua Blanca Fault, through rugged hills in the granitic and metamorphic rocks onto the flat El Rodeo surface. Miocene volcanic rocks form the steep mesas to the northeast. The Highway travels through rolling hills in the granitic and metamorphic rocks, then skirts a large flat alluviated area and approaches a low line of hills which mark the approximate trace of the San Miguel Fault Zone. The Highway then descends a gentle valley in the rugged tonalite hills along the trace of the San Miguel Fault Zone to San Salvador.*

San Salvador to Ensenada *- The Highway continues to follow a gentle valley along the San Miguel Fault Zone in the rugged granitic and metamorphic hills, then turns west and descends a relatively smooth rocky surface to the broad flat alluviated Ojos Negros Valley. It then passes one of the numerous isolated steep metamorphic hills which dot the valley and heads toward an escarpment of rugged gneiss hills with isolated granodiorite bodies. The Highway crosses the frontal fault and passes through rugged bouldery granodiorite hills to descend a steep narrow valley in the rugged hills of granitic and metamorphic rocks and climb a steep grade in rugged metasedimentary and metavolcanic hills. After going over a pass the Highway descends the narrow rugged Arroyo del Gallo in the metavolcanic, gabbro and tonalite hills. After climbing out of the arroyo at Piedras Gordas it follows a rolling ridge through tonalites, gabbros, and gneisses and finally descends a grade through rolling tonalite slopes to Ensenada.*

Follow the reverse of Log 14 for the 50 kms between San Felipe and the San Felipe Junction. If you are returning to Mexicali follow the reverse of Log 14 to Mexicali.

140 This is the junction of Highway 3 to Ensenada (195 Kilometers) and Tijuana (300 Kilometers). Turn to the left and begin to cross the peninsula. At this point the scarp of the Sierra San Pedro Martir and the peak of Picacho del Diablo are directly ahead. The Sierra Borrego hills are to the right; the Sierra San Felipe are the hills to the left. The Sierra Pintas are to the far right.

195 The kilometer markers now decrease to Ensenada. The crest of the Sierra San Pedro Martir (except for the peak of Picacho del Diablo) is very flat. This face of the range is actually a 3000 meter high scarp. Valle San Felipe in front is very close to sea level. The valley has over 2000 meters of alluvial fill making the offset on the fault in excess of 5000 meters (3 miles).

194 The highway makes several bends and heads directly toward Cerro El Borrego, the high peak of the Sierra San Felipe (slightly to the right of the highway) which is a granodiorite pluton. The lower grayer rocks to the right around it are prebatholithic carbonate sedimentary rocks. The slightly lower hill to the left of Cerro El Borrego is also granodiorite. Running off to the north, the hills are a mixture of granodiorite, tonalite, and prebatholithic carbonates until, at about 3 o'clock, they become the Miocene volcanic rocks of the Sierra Pinta. The vegetation is largely the same. As in other parts of Baja, sometimes it seems like it is all Ocotillo, other times all Smoke Brush or Mesquite in the washes. It basically depends

on what part of the slope they are on, what the substrate is, and at what angle the sun hits. It seems like the vegetation is constantly changing, however, it is just dominance that is changing. In the spring this section of desert is ablaze with the yellow flowers of the Brittlebush.

186 The highway again follows the elevated fan with washes on either side that are about 10-20 Meters below the fan. This whole area has been uplifted slightly and the washes are now regrading the fans to the new base level. Here and there the road drops down into one of these washes where the Mesquite and Smoke Trees become dominant. The little gray rounded balls are Burro Brush and Brittle-Bush which are more plentiful in the washes than up on the fans. The bright green plant along the road is Cheesebush.

179 Pass abreast of the Sierra San Felipe. There is a granodiorite hill to the right with a low skirt of metamorphic rocks around it. Notice the difference between the weathering characteristics of the Sierra San Felipe, which is very light, and the darker granodiorite cut by a number of small dikes which forms the low hills at about 2 o'clock.

The highway begins to pass by and through a series of hills to the left which are part of the old Pliocene alluvial fan which has been largely removed by the present drainage leaving isolated hills. In the Pliocene the top of these hills would have been the surface of the alluvial fan. Now the surface has been elevated and forms hills which are being dissected by the new stream drainages. In this desert terrain, you typically get severe erosion of older sedimentary rocks which are then remixed into the newer sediments and washed down into the washes.

173 Roadcut in Pliocene conglomeratic sedimentary rocks. Out of the valley into the higher elevation there is more Ocotillo, Spanish Bayonette, Teddy Bear Cholla, Creosote (replacing Cheesebush as a roadside opportunist), Barrel Cactus, and Brittle-Bush.

171 View south of Laguna Diablo and Valle San Felipe. Very thick Teddy Bear Cholla and Creosote form a very thick underbrush. Taller plants are Ocotillo and some Acacia.

169 The granodiorite hills are now next to the road. These rocks exhibit a prominent fracture pattern and are spheroidally weathered. They are cut by a light colored dike swarm.

165.5 The hill to the left, Cerro Coyote, is capped by a small patch of basalt.

164 Good view of Valle Santa Clara. Notice how very small the alluvial fans are on the face of the high range. Their small size, for such an elevated range, indicates very recent, very active faulting. Some of the fans to the south bear slight benches which were produced by faults cutting across the fans. At about 10 o'clock there are a series of dissected fans which have very obvious fault scarps on them.

On the floor of the playa lake there is a forest of Smoke Trees. Cheesebush is growing under the Smoke Trees. Cholla, Ocotillo, Creosote, Acacia with Mistletoe, Agave, Desert Mallow, and tufted grasses. Coyotes have been seen in this area.

154 As the highway enters San Matias Pass there is a gray granodioritic hill on the right. The small brown hill in front is gneiss cut by numerous dikes. The highway will pass through a roadcut in this gneiss, which shows the series of dikes. The highway then enters Cañon San Matias, proper, which is a very interesting low level pass through the range considering the impressive scarps on both sides. There is very little climbing, except what is needed to get up to the level of Valle Trinidad on the other side of the mountains.

DIKES CUTTING GNEISS

As the highway enters Cañon San Matias there is a whole series of sheet dikes. There seem to be, at times, more dikes than any other rock. The south side of the pass is largely gneiss cut by the dikes and the north side of the pass is largely granodiorite, which probably provided most of the dike material. There is some gneiss on both sides of the pass. The rocks on both sides have been mapped in detail. There is a series of K/Ar dates in the pass, too many to list.

This pass seems to be at the end of the Agua Blanca Fault Zone which was crossed near Ensenada. This fault zone is rather wide with quite a bit of movement on it and the pass is a logical place for it to go through. It now appears as though the fault simply ends here or continues through the pass as a subcrustal feature. This is unusual since it is a very strong feature with tens of kilometers of offset near Ensenada.

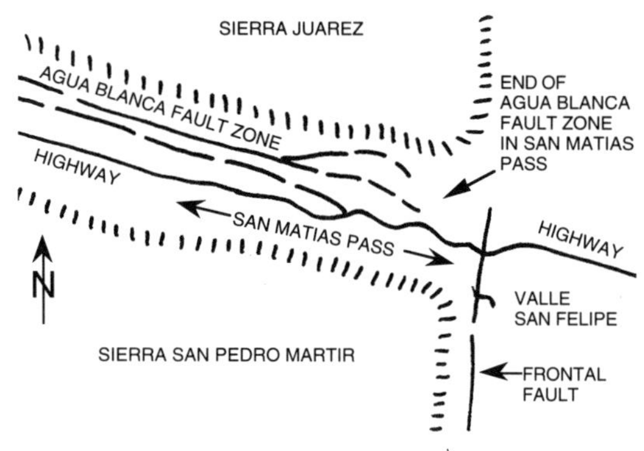

150.5 The roadcuts on the south side of the highway show the relationship between the metamorphic rocks, tonalites, and the dikes.

149 The dominant tall vegetation is Creosote with scattered Ocotillo. The north facing slopes have tall Ocotillo, Agave, Acacia, Creosote, Brittle-Bush, and Datillo. The south facing slopes have Brittle-Bush, small Ocotillo, and Creosote.

148 Below the highway, in the wash on the right, is the old road through this pass. It actually was a high speed road in this area. There are a lot of Barrel Cactus in the pass.

145 At Valle San Matias the Agua Blanca Fault Zone forms the north side of the pass. The straight-faced tonalite hill that comes down in a smooth slope is the fault scarp of the Agua Blanca Fault Zone which, here, shows a vertical component of normal movement.

The vegetation consists of Datillo, Pencil Cholla, Yucca, Creosote, Acacia and some Mistletoe. Honey Mesquite forms a dense almost forest-like area. There are Northern Harrier hawks and flocks of pigeons in this area.

The **Northern Harrier** or **Gavilán ratonero** has a distinctive white rump and owl-like facial disk. Their slim bodies are grayish above, mostly white below, and have black wing tips. Northern Harriers inhabit wetlands and open fields. Perching low and flying close to the ground, they search for mice, rats, and frogs.

138.4 Road to Mike's Sky Ranch and the Sierra San Pedro Martir. This road connects to the main highway Mexico 1 at San Telmo (2:140.9).

132 To the left there a swarm of dikes cutting the dark gray gneissic rocks. Ahead is the main area of Valle Trinidad. The hill to the right is basalt covering rhyolites and fluvial sedimentary rocks. The lower parts of the hills are lahars and fluvial sedimentary rocks. The upper part is basalt. Much of the geology is not obvious due to vegetational cover.

127 The highway follows directly along what is mapped as one of the traces of the Agua Blanca Fault Zone. A second trace follows along the base of the hills to the right. As the road bends, it comes directly onto the line of the fault, then bends away.

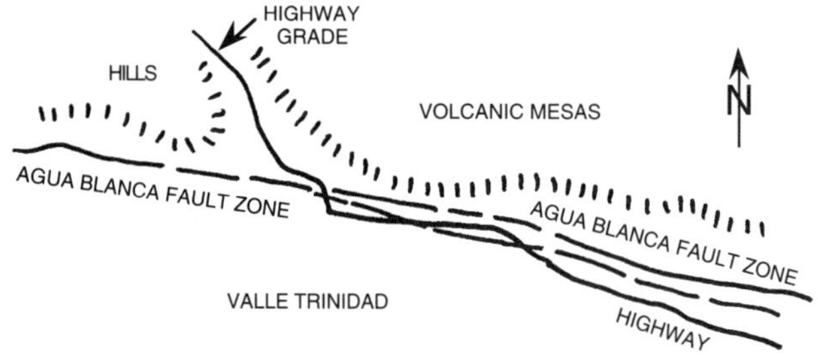

125 On the hillsides to the right are lots of Barrel Cactus up to a meter in height, all leaning to the

southeast. The hills are covered with Beavertail Cactus which is a sign of overgrazed disturbed soils. There is also Jumping Cholla, Creosote, and Yucca growing among the Beavertail. Creosote grows thickly in the washes.

121.3 Turnoff to Valle Trinidad is on one of the traces of the Agua Blanca Fault Zone. The hills to the west are composed of gneisses and tonalite. The road cuts up a small valley to climb from Valle Trinidad to cross the fault scarp onto the El Rodeo erosion surface.

OLD VALLE TRINIDAD GRADE

119 The old road up this grade was extremely rough. Early Baja travelers who went down it were often unable to return, and had to go out through San Felipe. This is an area of mixed rock with dikes cutting through tonalite. There are zenoliths in the tonalite and numerous faults and joints. Climbing the hill the Ocotillo is replaced by the Chaparral of the California Phytogeographic Region consisting of Manzanita, Toyon, Chemise, Oak trees, Rabbitbrush. (*See* 1:12)

114.5 Top of the grade. The view opens to the east of the high rhyolite capped mesas.

110 The road leaves a small hilly area and passes onto the El Rodeo erosion surface in an area of gneisses with some tonalite. This impressive flat erosion surface was cut during the Eocene. To the west at about 1 o'clock the two low hills are capped by gravels. These were filling part of a stream course flowing on the erosion surface during the cutting of the erosion surface. Now due to inverted relief, they exist as hills on the erosion surface. Since these conglomerates contain some of the 10 my old rhyolites, the streams must have been carrying the gravels sometime less than 10 million years ago. This gives a minimum rate of erosion of 100 ft./million years, or 1 mm/300 years, for this surface.

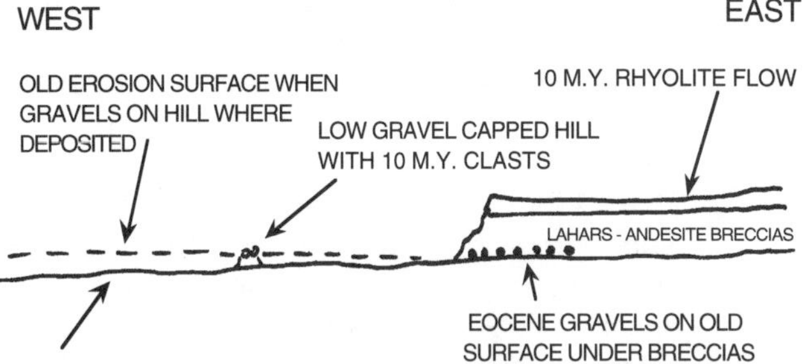

EROSION SURFACE NEAR LAGUNA HANSON

EL RODEO EROSION SURFACE FROM LOW GRAVEL HILL AT KM. 110

On top of the erosion surface there is a change in the vegetation: Junipers and Purple Sage are now dominant. This area looks like a Pinyon-Juniper pygmy coniferous forest without the Pinyon Pines. There are also Acacia, narrow-leafed Yuccas (*Yucca angosteda*). Beavertail Cactus is on the disturbed soil along the road.

107 The conglomerate capping the lava hill is part of an old river channel.

There is a view to the north of the very flat El Rodeo erosion surface which climbs gradually into the high peaks of the Sierra Juarez. The mountains do not look very high from this vantage point yet they are quite steep and rugged in some places. The erosion surface was developed when this area was closer to sea level. It has been elevated and tilted to the west (See 13:109).

103 Ejido Reforma. There are Meadowlarks, Ravens, and House Finches in this area.

102 The highway is on the flat erosion surface. To the west the hills which seem to rise from the edge of this surface are the coastal range of metamorphic mountains. To the east is the ring of mesas of rhyolite and basalt capped lahars.

99 Pinyon Pines now join the Junipers in the pygmy coniferous forest. This area used to be covered with Pinyon Pines and in some areas Ponderosa Pines. However, the gold mines of El Alamo consumed much of this forest for the charcoal to smelt the gold and timber to shore the mines.

97 The low range of hills directly ahead represent a horst between a pair of faults that are at a right angle to the Agua Blanca Fault and seem to offset one of the branches of the Agua Blanca Fault Zone. The pond is a sag pond developed as a result of the impounded drainage along the horst.

Mockingbirds and Scrub Jays are in this area.

91 The low hills to the left are covered by Miocene volcanics, The road is running approximately along a normal fault scarp, with schist on the right and tonalite on the left.

89 The road drops onto the flat, alluviated valley of Llano Colorado. Below the main hill in the far distance is the gold mining town of El Alamo. The main hill to the right of that hill is gabbro.

86.4 This turnoff leads to El Alamo (Cottonwood).

83 The road climbs into an area of granitic and metamorphic rocks. The rocks in the roadcuts are sheared, altered, and cut by numerous dikes. These dikes very typically contain sphene.

As the highway leaves the Llano Colorado it rises up onto the terrace again which is vegetated with the pygmy coniferous forest. Rabbitbrush, Yucca, Agave, Pinyon Pines, Juniper, Laurel Sumac, Toyon, and Manzanita are typical species.

81 To the right is what appears to be a reasonably straight line of low hills. These hills mark the trace of the San Miguel Fault Zone. In 1957 this fault zone was the location of a series of earthquakes. The road is still passing through schist.

76.5 The schistose nature of some of these rocks is very evident. The road leaves the flat surface and begins to descend a valley through tonalite.

74.5 Pino Solo area - In the old days here was one large Ponderosa Pine left in this area (Pino Solo). It was a giant survivor of what was here before the extensive logging claimed most of the trees in the area.

74 The spring to the right along the road is at the end of a fault line. The Broom Baccharis once again becomes the dominant species.

72.3 The small conical peak of Cerro Colorado comes into view to the left (1/2 Kilometer). It is tonalite which has weathered to a reddish color rather than the usual gray.

Quail, Mockingbirds and Scrub Jays are often seen along the road from Pino Solo to here.

69 The highway crests a small rise and begins descending into a little valley. Ahead is the extremely flat crest of the Sierra Juarez. The fault which fronts the Sierra Juarez into Ojos Negros Valley forms the jagged slopes directly ahead. The vegetation consists of stands of Broom Baccharis, Chemise, Purple Sage Brush, Willows, Manzanita, and low yellow flowered composites.

67 Occasionally there is a glimpse, in the wash on the left, of the old road. In the old days, this was a fine stretch of road because it was a long sandy

and smooth wash which allowed traveling at a pretty good speed. The road was lined with Broom Baccharis blocking the view of the road ahead, however, sound carried well. You would be traveling down the road at 30-40 mph, and meet someone who would be traveling up the road at a similar speed. At the last moment both of you would end up digging one wheel up out of the rut, touching a little brush, and passing each other with ease.

64 The darker hills without bouldery outcrops ahead and left are composed of gabbro.

62 San Salvador. Along this stretch of highway there are several ranchos, with Cottonwoods and Sycamores planted for windbreaks and shade.

59 A running stream bordered by large willows crosses the road. This stream is from a spring which is along one of the faults in this area. It is a nice rest stop. There may be some superstition surrounding this area. Years ago when we tried to camp in this area some of the locals coming through spoke of spirits and ghosts and "bad blood". Sangre de Cristo is a little village near here and we weren't sure if they just didn't want us to camp there or whether they were really afraid. The topographic map indicates that there was a graveyard in this area which may be the reason they spoke of ghosts, etc.

55 The view opens to the coastal range called Cerro el Encino Solo (hill of the single oak), of metavolcanic and granitic foothills west of a fault, and the very broad, flat, agricultural valley of Valle Ojos Negros. This valley is heavily alluviated and has extensive groundwater resources. The valley is bounded on the west side by a series of fault scarps and on the east side by an obscure fault, which has little topographic expression.

51 There is a riparian stream side community to the south of the road. There are many Cottonwoods with Mistletoe in them. The natural vegetation is mostly coastal scrub with Cat's Claw Acacia, Purple Sage Brush and lots of Mesquite.

49 The road passes into Valle Ojos Negros valley. A prominent granitic dike cuts across the road at the edge of a tonalite body. The hills to the west of the valley are variously composed of tonalite, schists and gneisses with some granodiorite forming very knobby outcrops. To the left ahead is the long linear scarp of one of the faults which forms the west side of the valley. The knobby appearing hill to the left ahead with all the boulders on it is a pod of granodiorite. Most of the rest of the hills are part of the prebatholithic gneiss (some schist) grading into what we call mixed plutonic rocks which are areas near the top of the batholith where the rocks are mixed together and it is difficult to separate the boundaries.

VALLE OJOS NEGROS

40 The highway takes a little bend to the left and passes by some very low gneissic hills.
In the agricultural fields in this area the mesquite look more like trees than shrubs. This is because the cattle love to eat the green shoots, exposing the trunk. There is also Mistletoe in the Mesquite.

Starlings and House Finches are seen here.

The **European Starling or Estornino**, like the Blackbird and Loggerhead Shrike, is a member of the Shrike family. The short-tailed Starlings are often confused with the unrelated, longer-tailed Blackbirds. Since their introduction into North America these gregarious, aggressive birds have become widespread and are seen roosting in Baja in a variety of habitats, including villages, agricultural fields, woodlands, telephone wires, fences, and open desert. Like the Gila Woodpecker, they nest in Cardons and other cacti.

Because of their ubiquitous occurrence, Starlings are considered "weed birds" and can be real pests often damaging cultivated vegetables and fruit orchards. They even do damage around feedlots where they consume and foul the feed of domestic cattle. Two field characteristics making Starlings easily identifiable are their yellow bills and glossy purplish-greenish tinted black head.

39.5 The road to Ojos Negros is to the right. The road to Ensenada continues straight to cut across a corner of the old highway. The hill ahead in the little gap is granodiorite as are the other hills to the left of the road. The hills, cut by white dikes, to the right of that are gneiss and plutonic rocks.

37 The highway crosses the frontal fault of the range near where a diagonal fault offsets it. The erosion along the diagonal fault line probably accounts for the low pass that the road utilizes.

36 The large spheroidally weathered bouldersindicate a general lack of, or widely spaced, jointing in the granodiorite.

35.5 The hills to the right are gneisses cut by light colored granitic dikes. The granodiorite is still on the left and comes to the road at about Km 35.

The hills are covered with typical Chaparral vegetation consisting of Buckwheat, Purple Sage Brush, and occasional Broom Baccharis with California Bays and Oaks in the stream bottoms.

34 Sharp curve. The roadcuts are on the contact between the granodiorite on one side and the gneisses on the other side. The road descends a grade through a series of mixed rocks which appear to be mostly tonalite cut by prominent, light-colored, granitic dikes.

29 The road climbs through prebatholithic metavolcanic rocks and then crosses a fault. Jimson Weed, which is a poisonous narcotic with white bell shaped flowers, is seen growing along the pavements edge from here to Ensenada.

26.2 Crest of the pass and a road to the left to Agua Caliente Hot Springs (7 Km) and San Carlos Hot Springs. The main road now descends various canyons and drainages to the coast at Ensenada.

25 There is a drastic change in the vegetation: An Oak woodland grows in the narrow valleys along the side of the road. Basically this is a riparian streamside community with Willows, Toyon, large Scrub Oak, and grasses. On the overgrazed hillsides the vegetation is characteristic of the Coastal Sage Scrub community with Yuccas, Chemise, "Witches Hair", Yarrow, Purple Sage Brush, and Black Sage.

21 Climb out of Arroyo del Gallo and into a gabbro body with its typical subdued topography and very few outcrops.

19 The highway rejoins the old road. This area of tonalite is known as "Piedras Gordas" or "fat rocks", mainly because of the large, rounded tonalite, and locally granodiorite, boulders. For the next several kilometers the road will pass through an area of mixed rocks consisting of gneisses and tonalites. Generally the tonalites have the bouldery outcrops and the granodiorites do not. Close to the coast the tonalites often weather to deep granitic soils because of the effects of moisture in the coastal weathering.

13 The road goes through a small pass for a view of the ocean, Bahia Todos Santos, Ensenada, and Islas de Todos Santos. As the road drops toward the coast, it passes through an area of tonalite which has been deeply weathered and is now covered by a rich soil.

11 The high hills to the right for the next kilometer are all metavolcanic rocks, as are the hills off to the left. The area near the road is a bowl in the less resistant tonalites.

3 Enter the main part of Ensenada.

0 The road crosses Highway 1 at the end of the bypass segment of Log 1

Log 13 - Tijuana to Mexicali [179 kms = 111 miles]

Tijuana to Tecate *- The Highway heads east from Tijuana on the elevated terraces of the Tijuana River with higher mesas of Tertiary marine sedimentary rocks north of the river and faulted rolling hills of Eocene sedimentary rocks to the south. The steep andesitic volcanic plug of Cerro Colorado dominates the north mesa. The Highway climbs a metavolcanic hill to cross Presa Rodriguez at a narrows in the metavolcanic rocks. The Highway then follows the marine and fluvial terraces on the south side of a gentle valley bounded by high rugged metavolcanic hills. The north side of the valley is a rolling terrace underlain by marine and fluvial sedimentary rocks and dominated by steep, conical, Pliocene andesite plugs. The Highway then crosses a belt of rugged, metavolcanic hills and continues through bouldery granodiorite hills with numerous gabbro pods. It crosses over a railroad and through an area of steep tonalite hills, finally reaching Tecate.*

Tecate to La Hechicera *- East of Tecate the Highway passes through steep hills of gabbro, follows a steep sided canyon between tonalite and granodiorite, and then through the granodiorite to emerge on a flat elevated erosion surface developed on tonalite. The Highway travels for many kilometers across this surface with views to the north of the Laguna Mountains and to the south of flat conglomerate capped hills.*

La Hechicera to La Rumarosa *- The Highway enters a relatively flat area of rolling gneiss and schist hills locally cut by pegmatites. Between El Condor and La Rumarosa, conglomerates capping hills are exposed on both sides of an alluviated valley. Isolated hills of marble in the rolling hills of granitic rocks are being mined on both sides of the Highway near La Rumarosa.*

La Rumarosa to Mexicali *- The Highway begins to descend the steep gulf escarpment and drop through bouldery outcrops of granodiorite and tonalite and then through gneiss and schist cut by dikes. At the base of the escarpment the Highway crosses the frontal fault and descends the alluvial fan to reach the rugged tonalite ridge of the Pinto Mountains. The Highway then crosses a fault in a wash and climbs onto the uplifted older fan surface to continue its gentle descent on the fans toward Laguna Salada in view to the right. The Highway crosses the active Laguna Salada Fault Zone and a low pass in the northern end of the Sierra de los Cucapas, a rugged complex area of granitic, metamorphic, and Tertiary sedimentary rocks. Signal Mountain is the high rugged tonalite peak to the left. Then the Highway descends gentle fans to the Colorado River Delta area of the Mexicali Valley.*

179 Move into the right two lanes as you cross the bridge and continue straight along the divided avenue heading eastward (Mexico 2) to follow the Tijuana River drainage with Pleistocene terraces on either side. The terraces on the right (southwest) have been disrupted by faulting. Turn right at about the third or fourth monument, then left on Agua Caliente.

176 Greyhound dogs run daily at the Agua Caliente racetrack. This racetrack was built on the site of the Agua Caliente Hot Springs on the Agua Caliente Fault Zone.

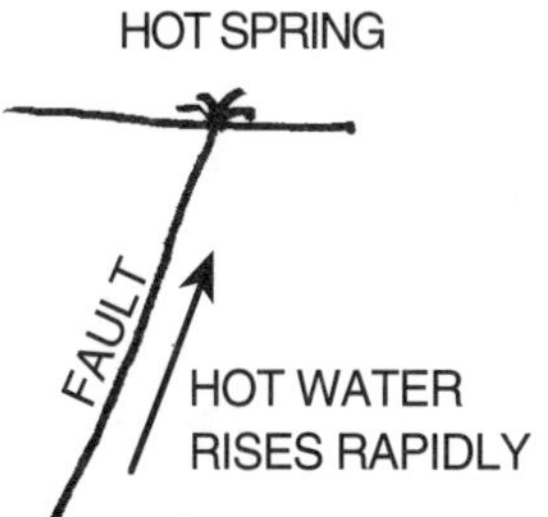

Hot Springs: A hot springs is at least 10° C warmer than the mean air temperature. Hot springs do not normally require abnormal heat sources. The earth's temperature increases about 1.8^{o} Centigrade per 100 meters of depth. Subsurface ground waters are heated to the temperature of the surrounding rock so the 10° warmer water need only come from about 550 meters below the surface. Faults provide an avenue for the rapid rise of hot ground water to the surface resulting in hot springs. Most of the 1000 hot springs in the western United States are along faults. Some such as Yellowstone and Lassen are near volcanic centers and are due to near surface magma.

NOTE: Kilometer markings are few and far between along this stretch of the highway so watch for them carefully in order to keep up with this part of the log. You may need to orient yourself at Presa Rodriguez

The highway proceeds southeastward for about 10 Kilometers paralleling the drainage of and traveling over the low river terraces of the Rio de las Palmas. The Rio de las Palmas roughly follows the path of the ancestral Las Palmas River that was Eocene in age. The high terraces on the left are part of the mesas which stretch from the San Diego area into the eastern Tijuana region. They are Pleistocene material underlain by Miocene and Pliocene sedimentary rock.

175 The highway passes a monument in the center of the road which is a representation of the Aztec viaduct in Mexico City. There are a number of similar monuments in other places in Mexico.

167 This is the first kilometer sign seen on Mexico 2 since leaving Mexico 1.

166.5 The highway makes a bend and begins the climb up onto the metavolcanic hills. The red double-peaked mountain to the left is Cerro Colorado, a Pliocene andesite plug. This is one of a number of such plugs which stretch east toward Tecate. To the left of Cerro Colorado is the lower Cerro San Isidro, another Pliocene plug. The plug closest to the U.S. border is Cerro la Avena. Otay Mountain, part of the San Isidro Mountains in the U.S., is the northernmost peak.

VOLCANIC PLUGS are cylindrical masses of rock sealing the vents and conduits of volcanoes which are exposed as the more erodible surrounding rock is removed.

The hills to the right, in the sedimentary rocks of the La Mesa region, were deposited in the delta of a large Eocene river. The high hills to the left beyond the mesa are metavolcanic.

165.4 The highway crosses Presa Rodriguez, a concrete dam 1,935 feet long, that was built in a narrow gorge between metavolcanic hills. The metavolcanics are exposed along the road for the next several kilometers.

The turnout on the left, as you cross the dam, is a good place from which to view the chaparral vegetation and the dam. The vegetation in this area is very sparse and is typical chaparral species (*See* 1:14.8 consisting of Black Sage, Chemise, low annuals, Cheesebush, Broom Baccharis, and Buckwheat. After rain there may be lots of low green herbage growing on the normally bare metavolcanic soils of this area.

164.6 The lake behind Presa Rodriguez is a reservoir supplying water for Tijuana. This reservoir is built along a possible fault line which is probably responsible for the alignment of the Tijuana River drainage. The low hills on the west bank of the reservoir are part of the Eocene sedimentary rocks. Ahead to the right, the higher hills are metavolcanic rocks. Nearly all of the rock in view ahead, as the highway makes the second bend on the dam, is metavolcanic rock.

164 Cerro las Abejas is the sharp triangular shaped volcanic plug on the left. The white flat region west of Cerro las Abejas is called Lomas Blancas.

163.5 The left fork in the highway returns to Tijuana on the north side of the Rio. Stay to the right to continue eastward to Mexicali.

161.5 The two andesite plugs visible to the left are Cerro Colorado and Cerro las Abejas. The highway follows a poorly defined valley in terrace material with metavolcanic rocks on the right and a terrace developed on Eocene? and Pliocene sedimentary rocks to the left.

161 The highway passes through a mass of granitic rocks.

160.2 This roadcut exposes Pliocene rocks.

156 The low rolling hills to the left are middle Tertiary sedimentary rocks.

153 The conical peak (Cerro La Posta) in front on the right is another andesite plug with Tertiary terraces developed on its sides. Prebatholithic metavolcanic rocks are now exposed on both sides of the road. The high hills to the right are metavolcanic and metasedimentary rocks.

The vegetation covering the hills in this region are, low chaparral species dominated by Sagebrush, Wild Buckwheat, Black Sage, White Sage, and various low annual herbage.

152 The level of the high terrace ahead to the left is the same as that of the andesite plug mentioned above (Km 153).

150 A marcadum ("asphalt") plant is located on the right side of the highway.

MARCADUM: Asphalt (or mineral pitch) is a brownish-black **solid** or semisolid mixture of bitumens (hydrocarbons obtained by distillation from coal or petroleum) obtained from native deposits or as a petroleum by-product, used in paving, roofing, and waterproofing. Asphalt is used in the U.S.A. for paving highways. In Baja, Marcadum is usually used as a paving material instead of asphalt. Marcadum is a naturally occurring **liquid** petroleum hydrocarbon used for paving, roofing, and waterproofing.

148 The vegetation growing in this wash are those typical of riparian stream side floral communities. The predominant large plants of the wash are Broom Baccharis, Mesquite, and Tamarisk. The Tamarisk is an introduced European species which has become naturalized in North America.

143 The highway enters an area of mixed granitic rocks and darker diorite cut by dikes with masses of bouldery tonalite on the hills.

140.6 The highway crosses over the railroad tracks of the Tijuana and Tecate Railroad near Los Lauelas. It used to be called the San Diego and Arizona Eastern Railroad (S.D. & A.E.). A trestle was washed out in Carrizo Gorge near Jacumba in the late 1970's and that section of the line has been out of service ever since. The San Diego part of the line is now the Tijuana Trolley.

139 The highway now passes through granodiorite into tonalite. The clay pits of the tile and brick operations of Tecate have been developed in the gabbro soils of this area.

136.5 Flat-bottomed native Oaks grow in this area. Cattle love to browse the oak foliage and so the trees are "pruned" as high as their necks can reach resulting in "flat-bottomed" trees. Trees growing where cattle do not browse are shaggy bottomed with foliage reaching to the ground.

136 The Rancho La Puerta is a resort and spa.

The boulders of this area are composed of tonalite. The high hill to the left with roadcuts in it is Tecate Peak which is on the U.S. side of the border in the San Isidro Mountains.

133 **TECATE:** The small town of Tecate (elevation 1,690') is located in a bowl-shaped valley surrounded by tonalite boulder covered hills. Tecate was established in the 19th century as an agricultural center because of its abundant water and fertile soil. Although home to both the Carta Blanca and Tecate beer breweries, agriculture remains the mainstay of Tecate's economy. The primary crops produced here are grapes, grain, and olives. The Zocolo ("village square") is shaded by Palm and Box Elder Trees.

131.5 Just east of downtown Tecate is the intersection of Mexico 2 (Avenida Juarez) and Mexico 3 (Calle Ortiz Rubio). Mexico 3 (Log 12) heads south-westward 107 Kilometers and connects with Mexico 1D on the Pacific coast at El Sauzal, 1.5 Kilometers southeast of San Miguel (Log 1:101.3). Highway 2 continues eastward and crosses over the tracks of the old San Diego and Arizona Eastern Railroad east of Tecate. The highway then follows the railroad eastward up a canyon.

129 Cerro la Panocha is the small round, gabbro hill that looks like the raw brown sugar (panocha) cones produced as the syrup of sugar cane is poured into cone-shaped forms. Cerro la Panocha straddles the border - its north slope is in the U.S. and its south slope is in Baja.

128 The highway follows a gabbro body on the left. The hills to the right are granodiorite which weathers into bouldery hills in this part of the Peninsular Ranges. To the left of the highway is typical riparian vegetation.

123 The railroad tracks of the S.D. & A.E. can be seen, across the valley to the left, running along the hills. Occasionally they are hidden from view by the vegetation and by the roadcuts. The tracks turn left and head north across the U.S. border.

120 San Pablo. Directly ahead to the left of the highway is Cerro la Rosa de Castilla.

119.5 The vegetation of this area is typical of the chaparral plant community and is predominated by sages and Buckwheat.

119 The Rancho Rosa de Castilla is located in a beautiful oak woodland area. For the next 6 Kilometers the highway passes in and out of several oak woodland valleys.

OAK WOODLANDS The three most common oaks of Baja Norte are Coast Live Oak (*Quercus agrifolia*), Scrub Oak (*Q. dumosa*), and Canyon Oak (*Q. chrysolepis*). These three oaks can be identified by studying the following chart.

characteristics	*Q. agrifolia*	*Q. dumosa*	*Q. chrysolepis*
evergreen	yes	yes	yes
height	10-25 m	5 m	6-20 m
growth habit	large, broad-crowned trees	shrub often forming dense thickets	large tree forming oak woodlands below 600
usual habitat	below 1000 m in coastal valleys and foothills bordering Chaparral in open grasslands	in Chaparral communities	in canyons, on north facing slopes, and in Chaparral
leaf	shiny, dark green,	shiny, gray-green,	oblong or ovoid, bluish-green
morphology	oval upper surface; hairy at vein on under sides of leaf; leaf margin toothed; up to 6 cm long	curled leaves; under surface covered with brownish hairs; leaf margin toothed; 1-3 cm long	green above; covered with a yellow or silver powder below; leaf margins toothed; 2-6 cm long
acorn	slender; pointed; 2.5-3.5 cm long		oblong-ovoid with turban-shaped cups; 2-6 cm long
bark	gray with broad checked ridges		smooth to scaly gray

Oak woodlands are in danger in both Californias wherever livestock are pastured because sheep and cattle eat young seedlings before they can grow to mature trees. Therefore, when the old, mature oaks die, oaks will disappear from grazing areas. Urban development also significantly reduces the acreages of oak woodlands.

118.5 The massive granodiorite cliff ahead is Cerro Rosa de Castilla. With a little imagination it looks something like Half Dome in Yosemite Valley.

118 Another oak woodland is growing in this valley

117 The hills are covered with bouldery outcrops and true Chaparral vegetation. The highway passes through a series of broad flat valleys in the granodiorite.

111 The highway climbs a long grade through diorite outcrops cut by light-colored granitic dikes. For several kilometers the highway passes through metamorphic rocks of the Julian schist which are also cut by quartz dikes.

110 The dominant plant species is Broom Baccharis which is growing in almost pure stands. It resembles a broom.

109 As the highway reaches the top of a long grade it begins to cross a broad flat erosion surface developed on the La Posta quartz diorite.

EROSION SURFACES: This surface was developed during the late Cretaceous and Early Tertiary when this area was closer to base level. It is the effect of prolonged weathering in a more humid climate. The flat-even topped hills on the skyline to the southeast are capped with Eocene conglomerates. These conglomerates were deposited by a large river which flowed across this area during the Eocene. The source of this river was near Nogales in South-Central Arizona. It flowed across the area of the Gulf of California which did not exist at that time. The delta of this Eocene river is in the Tijuana area.

When streams are near their base level they cannot erode downward. Instead they cut laterally and tend to form a flat plane. If this surface is not uplifted it will remain a flat surface. Even when uplifted, the streams on the surface do not have any power until they reach the edge and begin downcutting. Over millions of years, weathering of the surface rocks and the carrying away of the finer material by the low-powered streams has lowered this surface about 1 mm/100 years. The conglomerates were not easily removed and, thus, became resistant ridges above the surface.

GRAVEL RIDGE ON EROSION SURFACE

An interesting point is that once an area is left above the surface it tends to remain drier and not weather as rapidly as the rocks on the flats. This slows the rate of erosion on the hill. As a result the hills tend to become more pronounced and higher above the surface with time

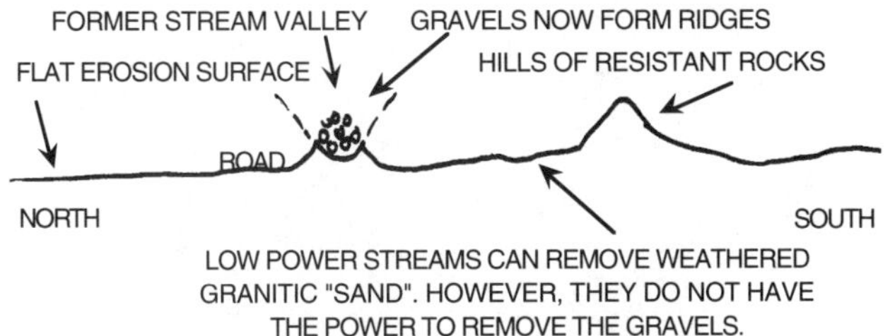

105.5 This turnoff leads southward to El Compadre and Valle Ojos Negros ("black eyes") and meets Mexico 3 at 12:55.3.

104.5 This side road leads to Rancho Jacomun. The highway passes through another oak woodland.

In this area it is possible to see the pipe which brings water for irrigation to this region of Baja, via an aqueduct, from the Colorado River.

103 There are large boulders covered with lichen here. Lichens are lower plants made up of single-celled algae living inside the tissues of a fungus. These two plants live in a mutualistic symbiotic relationship in which both benefit. Organic acids secreted by lichens degrade rock into soil preparing the way for the establishment of higher plant forms.

Red-stemmed Buckwheats are growing in the disturbed soils along the highway shoulders. The seeds of Buckwheats were ground into a meal and baked into cakes or made into a mush by native Baja Indians.

100 There is a large bend in the highway as it turns to the south and passes through the small village of Hechicera.

98.5 The small village of Luis Echeverria Alvarez (El Hango) is to the right of the highway.

97.5 The gray bank low on the hills to the left is a rock quarry in the La Posta quartz diorite.

97 The distant Laguna Mountains of San Diego County can be seen from here. Their flat concordant summits are part of the same erosion surface at about 1200 Meters higher elevation. By a round about route, the Laguna erosion surface descends gradually to the level of the surface near the road.

93 The stringers of white boulders exposed in the hills for the next 4 miles are pegmatite dikes.

89 This is a good turnout to view the vegetation of this region such as Broom Baccharis with Laurel Sumac, and Black and White Sage. Scrub Jays may also be seen in the area.

87 The flat-lying pegmatite dikes on the hill, under the power lines, have yielded gem Topaz, Beryl, Garnet, Tourmaline, and Smoky Quartz crystals. Pegmatite dikes such as this one are abundant in San Diego County. Here the highway passes through exposures of thinly-bedded prebatholithic gneisses and schists assigned to the Julian Schist. These metamorphic rocks are probably Paleozoic in age and represent the remnants of the rocks into which the granitic rocks and pegmatite dikes were intruded.

83.5 Good examples of Julian schist are exposed in this roadcut.

PEGMATITE DIKES

82.8 This turnoff leads to the small settlement of El Condor located on the general level of the high erosion surface. The road south from El Condor connects up with the main La Rumarosa-Laguna Hansen road after about 15 Kilometers. This was used as a cutoff by early Baja travelers heading to Laguna Hansen in the Sierra Juarez.

The hills ahead to the left and to the right in the middle distance are covered with Eocene conglomerates discussed at Kilometer 109.

81 Just east of El Condor the highway enters a Pinyon-Juniper coniferous forest. Scrub Oak, Broom Baccharis, Chemise, Yucca, Black Sage, and White Sage also grow in association with this coniferous forest. Ravens may be seen in this area.

THE PYGMY-CONIFEROUS FOREST: In Northern Baja the California Juniper (*Juniperus californica*) is associated with the Pinyon Pine (*Pinus quadrifolia* and/or *P. monophylla*). This association forms the Pinyon-Juniper community, which, due to the short trunks, is commonly called the "pygmy" coniferous forest.

The leaves of Junipers have been reduced to bright green, imbricated (overlapping like tiles of a roof) scales that closely clothe the smaller branches. They are coated with a waxy cuticle that helps keep moisture in. Another adaptation reducing water loss are the small, dry, berrylike fruits with their protective waxy bloom. The shaggy ash-gray bark provides insulation from the desert heat, keeping tissue temperatures down, another water-conserving adaptation.

The Pinyon Pine grows taller than the Juniper. The leaf stoma of the Pinyon are found in the bottom of pits in the needles, reducing water loss. The pits appear as small white stripes on the needles. The Pinyon has a fungus growing on its host's roots (a mycorrhizal mutualistic association). The fungal symbiont provides water for the tree and protects the roots from disease; the host Pinyon provides sugar for the non-photosynthetic fungus which cannot produce its own. A walk into this pygmy forest will quickly reveal the identity of Pinyon Pine and the Juniper.

Both Pinyon Pines and Junipers have been and are still useful to man and animals in Baja. People still collect the protein rich seeds, or Pinyon nuts for food. Many of Baja's birds, rodents, and deer also rely on the Pinyon nuts for food. The nuts are produced on the sporophylls (seed leaves) of the large female pine cones. Male cones are very small, producing only pollen. Today, the wood of the Juniper is primarily used to make fence posts in Baja. The cones of *Juniperus communis* are soaked in alcohol to leach the oils from the seeds and cone tissues which are used to give gin its characteristic flavor.

PINYON JUNIPER WOODLAND VOLCANIC CAPPED PEAK ON THE EROSION SURFACE

Both the Juniper and Pinyon Pine are especially plentiful on the dry slopes of the Sierra Juarez and Sierra San Pedro Martir below 5,000 feet in the Upper Sonoran Life Zone. Overgrazing in northern Baja is allowing them to increase in abundance and extend their domain into places formerly occupied by grasses thus reducing the available range land and necessitating increased use of cultivated forage and fodder. While reducing range lands, their increase is helping native hoofed-browsers like Deer and Antelope which feed on the twigs and foliage of both conifers. In addition to their wildlife food value, they also provide important protective and nesting cover for numerous bird species in Baja like Robins, Sparrows, Mockingbirds, and Warblers.

76.5 The hills on both sides of the road are composed of conglomerates laid down in a Eocene river. 30 Meters east of the 76 Kilometer mark there is a dirt road to the left which can be followed for 200 Meters northward to exposures of the Eocene conglomerates. These conglomerates are poor in the usually present exotic meta-rhyolites due to the mixing with materials of another Eocene stream which contained no meta-rhyolite clasts. The highway is sitting on a metamorphic schist and gneiss basement at this point.

73 Marble quarried from the roadside quarry on the hill to the left and others in the area is used as decorative rock and to make cement. Several small caves decorated with stalactites have been found in the marble. This is of concern to the miners as caves reduce their marble reserves.

The thick Pinyon-Juniper forest growing in this region is made quite scenic by granodiorite bouldery outcrops. This area is reminiscent of the Catavina boulder fields (*See* 3:153.5) except the boulders here are smaller and more numerous.

67.5 La Rumarosa is a summer resort for residents of Mexicali. The fault scarp which is much steeper and more evident to the south is not as steep here due to exposures of the more easily eroded schists. A dirt road leads southeastward from La Rumarosa, into the Sierra Juarez, to the Parque Nacional Constitucion 1857 and Laguna Hanson. Upon leaving La Rumarosa, the highway begins its steep, winding drop down the Cuesta La Rumarosa into the desert along the eastern gulf drainage of the Sierra de Juarez. The rocks exposed at the top of the grade are granodiorite. Further down the grade tonalites and finally schists are exposed.

65 The highway crosses through an area of spheroidally weathering granodiorite. (*See* 3:94.) Notice that the flat westward sloping erosion surface stretches to the crest of the range. The steeper gradient Gulf streams are rapidly cutting westward into the scarp. This descent of the Cuesta La Rumarosa (Cantu Grade) is the steepest and most dangerous paved road in Baja.

On the slopes of the eastern drainage of the Sierra Juarez, the granodiorite boulders are intermixed with beautiful Pinyon Pines, Junipers, Manzanita, White and Black Sage.

RIFT VALLEYS AND THE GULF OF CALIFORNIA: The East Pacific Rise is passing under the continent and forming the Gulf of California and the Great Basin of the southwestern U.S. and northern Mexico (*See* introductory geology). This results in a series of mountains (horsts) and valleys (grabens) bounded by faults. As these mountains are eroded, broad alluvial aprons (bahadas) stretch out from the hills into playa lakes. Eventually the rise will pull Baja away from the rest of the continent along the San Andreas fault. Offset on the San Andreas fault (in the far distance) averages 2 inches per year. At that rate San Francisco will appear on the other side of the fault in about 30 million years. 2"/year = 16 miles/million years = 480 miles/30 million years

62 There is a view to the left into a bouldery valley with a small stream. The pipe with two green surge towers is part of the pipeline bringing water from the Colorado River to Tijuana. The desert to the northwest in California is called the Yuha Desert. In the far distance West Mesa is visible. Directly to the left, just barely beyond the edge of the hills is the U.S. town of Ocotillo. To the right of Ocotillo, in the far distance, the Salton Sea is visible. Further to the right the white spot in the nearer distance is Plaster City. Gypsum mined from Split Mountain in the U.S. is hauled to Plaster City via the railroad where it is processed into wall board primarily used in residential construction. This railroad is part of the San Diego and Arizona Eastern Railroad. Although this railroad is still utilized in the Imperial Valley, it no longer goes up the grade. Just barely in view to the right is Signal Mountain or Cerro la Centinela, a high peak in the Sierra Cucapa. The hills in the far distance, northeast of Split Mountain are the Chocolate Mountains.

61.5 This turnout on the edge of the granodiorite pluton offers a view of the San Felipe Desert. There are several other turnouts which offer good views.

59.5 First view of Laguna Salada.

58.5 A dark gray to black basalt dike cuts the granodiorite in this roadcut.

57 As the highway comes around a corner a bedrock landslide is visible at Km. 56.8.

56.7 The buildings on the ridge to the left are part of a pumping plant for the pipeline that pumps water up and over the Sierra Juarez westward to Tijuana. The green surge towers release the air that gets trapped in the water preventing large air bubbles from building up and blocking the movement of the water as it descends along the western slopes of the Sierra Juarez.

55 As the highway nears the bottom of the Cuesta La Rumarosa grade, in view to the left are Laguna Salada (foreground), Cerro Colorado (the dark linear ridge in the middle-ground), and the Sierra Cucapas with Cerro la Centinela (background). Cerro Colorado is composed of granitic rocks covered with a dark-colored desert varnish.

The Sierra Cucapa are a lighter reddish-brown and lie in the background on the far side of Laguna Salada. They are also covered by a patina of desert varnish but not as dark as the surface of Cerro Colorado.

DESERT VARNISH: A minor weathering feature of the desert is a thin, shiny, reddish-brown to blackish coating called desert varnish that occurs on some desert rocks. This shiny blackish-brown stain is only one or two micrometers thick and appears to be an amorphous gel, rich in silica and alumina that takes its color from unusually high concentrations of iron and manganese. The chemicals of the varnish originated from atmospheric dust, the stones themselves, and underlying soil and chemicals dissolved in films of moisture formed on rock surfaces by rain, fog, or morning dew.

As moisture evaporates from warming rock surfaces it leaves ions of the dissolved chemicals behind as a coating that gradually builds up over hundreds of years, increasing the thickness of the varnish. The rate of varnish formation is extremely slow.

Desert varnish occurs on the upper surfaces of rocks on alluvial fans and the mosaic pebbles of desert pavement. Similar stains are widely found around seeps, especially as dripping, large, black streaks on canyon walls. The undersides of many varnished rocks are orange (iron rich) while only their sides and tops are blackish-brown (manganese rich). Since iron is less soluble than manganese it is left behind under the rocks as the more soluble manganese is drawn upward to concentrate on the exposed sides and upper surfaces.

53 The highway passes through metamorphic gneiss and schist of the Julian Schist with pods of granodiorite and tonalite mixed in here and there.

The hills of this area are vegetated with Yucca, Teddy Bear Cholla, Palo Verde and Ocotillo. This area is a good place to observe vegetational differences between north and south facing slopes.

49.7 A pegmatitic granitic dike cuts the metamorphic rocks in the roadcut on the left.

48 To the right the Rio Agua Grande runs through the Canon los Llanos. There is some water as evidenced by the row of vegetation following the meander. The flora of this area is typical of a Wash Woodland vegetated by Willows, Stinging Nettle, Acacia, and Creosote Bush.

44 This is a good turnout for viewing the vegetation of the eastern drainage of the Sierra Juarez. Creosote Bush, Bursage, Brittle-Bush, Palo Verde, Tamarisk, Cat's Claw Acacia, Ocotillo, and the southward leaning barrel cactus are all visible. Specimens of *Pachycormus discolor*, the Elephant Tree, grow here and northward into the Anza-Borrego Desert.

43.5 The highway crosses the Rio Agua Grande. The predominant trees growing along the Rio Agua Grande are Smoke Trees and *Acacia* parasitized by mistletoe. Both trees are characteristic and indicator species of Wash Woodlands.

WASH WOODLANDS: Because of the limitations of water supply in the area occupied by Creosote Bush and Bursage, the water courses, which are dry most of the year, support a characteristic flora that takes advantage of the relatively abundant supply of water during rainy periods of the winter in Northern Baja. This region of a wash and its characteristic flora is a Wash Woodland. This is a separate plant community which does not belong to the surrounding arid Creosote Bush Scrub plant community. The flora of the Wash Woodland is dominated by Palo Verde, Smoke Tree, Cat's-Claw Acacia, and Mesquite.

SMOKE TREE

The seeds of many Wash Woodland species are covered by a very hard (sclerophyllous) seed coat (integument) and will not germinate no matter how long the seed is soaked in water unless the seed coat is broken (scratched through) by a process called scarification. That is, it is necessary to scratch the seed coat, allowing water to enter the seed and initiate seed germination. Scarification is accomplished by the grinding action of sand and rocks in the floods occurring periodically in the washes. This also provides the germinating seedlings with water which will supply their growth requirements during the first few weeks of germination. Flash floods also serve to dispense the seeds. Like many desert perennials, seedlings of Wash Woodland trees produce only two or three leaves immediately after germination and seemingly become dormant. However, these plants are devoting their energies to developing extensive, deep root system that will enable then to survive long after the moisture from the infrequent floods have dissipated. Plants developing deep root systems enabling them to tap underground water sources are known as phreatophytes (phreato = well; phyte = plant).

42 The highway descends on the alluvial fan surface at the base of the Sierra Juarez.

The vegetation growing on the alluvial fan surface consists mainly of Ocotillo, Brittle-Bush, Cholla, and Creosote Bush.

37 The highway crosses another of the meanders of Rio Agua Grande developed along a fault and climbs onto an elevated part of the fan surface. This fault attests to the tectonic activity which continues today in the Imperial Valley and along the northern Gulf of California. The darker, desert varnished hills to the north along the fault are the Pinto Mountains. This area has been a popular collecting area for the colorful "Pinto Mountain Rhyolite" and petrified wood which is often polished by the wind. Be careful not to stray across the border if you explore this area. The road to the south skirts the foot of the Sierra de Juarez and eventually reaches a beautiful palm oasis at Canon Virgin de Guadalupe (a better road is located at Km 27.5).

34 The prominent peak to the left ahead is Signal Mountain (Cerro la Centinela), a high granitic peak at the north end of the Sierra Cucapa. The low hills to the left consist mostly of Tertiary sedimentary rocks uplifted along the active Laguna Salada Fault Zone.

30 The view to the rear is of the impressive eastern granitic scarp of the Sierra Juarez.

27.5 The road to the right goes down the Laguna Salada to Canada Cantu de las Palmas and Canon Virgin de Guadalupe.

Laguna Salada is a rift basin bounded by faults. It will continue to enlarge and will eventually become a permanent part of the Gulf of California. Laguna Salada was nearly dry in the early 60's, receiving only run off from the local mountains. It has had considerable amounts of water in it in recent years due to major storm surges in the Gulf of California, during high tides, which pushed large amounts of sea water into this basin filling it to the brim. It is supertidal at the south end and connects with the Gulf of California during very wet periods (*See* 14:73). The Laguna Salada Fault Zone passes close to this junction.

25.5 To the left a large dune has developed in the lee of one of the nearby hills.

The road to the right leads to El Centinela campground located on the beach of Laguna Salada.

The trees on this beach were underwater as evidenced by their barnacle encusted trunks. Barnacles can be seen as high as 3.5 Meters off the ground. Both the water and the soils of the laguna are very salty. Salt crystals "growing" on the surface of the shore soils are formed by efflorescence as the salt-laden water evaporates. This beach is muddy and the laguna is shallow.

LAGUNA SALADA

23 The highway makes a sharp bend and winds its way through a small pass which marks a small fault and a horst of metamorphic rocks. The hill in front is highly desert varnished. As the road turns to the right, the vertical striations on a fault plane are visible ahead at road level.

Adjacent to the highway the vegetation consists of tall *Acacia*, smaller Smoke Trees, Creosote Bush, Bursage, and scattered Tamarisk trees.

20 The highway now descends east-sloping alluvial fans into the fertile Mexicali Valley.

19 The *Acacia* in this region is parasitized by Desert Mistletoe (*Phoradendron californium*).

MISTLETOE is a woody, perennial evergreen parasite that steals sugar from its host by way of a modified stem called a haustoria. The sticky seeds are disseminated on the feet and bills of birds that eat the pearly-pink mistletoe berries.

15 This area was formed by the Colorado River as it built a delta across the northern end of the Gulf of California. The Imperial Valley in California is a segment of the gulf that was cutoff and isolated by the Colorado River Delta. Indio, California, is 11 feet below sea level. The low point of the delta between the Imperial Valley and the Gulf of California is only 35 feet above sea level. As the delta was built the Colorado River sometimes flowed into the now isolated Imperial Valley. This created a series of lakes alternating with dry periods as the Colorado River alternately flowed into the basin and the Gulf of California. In the Imperial Valley the high level is marked by algal limestone along the shoreline of the ancient fresh water Lake Cahuilla at Travertine Point. However, the present Salton Sea is man-made. It is the result of an error in judgement in 1906 which allowed the full flow of the Colorado River into the valley for over a year.

4 A bull ring is to the right of the highway. East of the bull ring Date Palms are being cultivated and Tamarisk trees have been planted for windbreaks.

0 The highway diverges at a "Y" fork. The right fork is the continuation of Mexico 2 which connects with Mexico 5 in another 8 Kms. The left fork leads into Mexicali.

MEXICALI, the state capital of Baja California, is the second largest city in Baja. Agriculture is the backbone of Mexicali's economy. Irrigation water for the Mexicali valley comes from the Morelos Dam, on the Colorado River just south of the Mexico/U.S. Border. Baja's first brewery, the Mexicali Brewery and tasting room, are also found here.

The paleontology, archaeology, and mission exhibits of the Museo Regional Universidad de Baja are well worth a visit.

0 Mexicali is to the left and San Felipe to the right on Mexico 5.

Log 14 - Mexicali to San Felipe [190 kms = 118 miles]

***Mexicali to the Sierra Pinta** - This segment of the Highway heads south across the flat delta sands of the Colorado River, crosses the Cerro Prieto Fault Zone, and climbs onto, and begins to follow, the undulating alluvial fans of the Sierra de los Cucapas.*

The high, rugged Cucapas to the west are uplifted fault blocks of granitic and metamorphic basement uplifted during the opening of the Gulf of California. The basalt cinder cone of Cerro Prieto and its geothermal area are to the east. South of El Tare junction the Highway crosses a delta mudflat where a low pass in the metamorphic rocks denotes the start of the almost continuous rugged granitic and metamorphic Sierra Mayor. To the left is the delta region with the Rio Hardy and the Colorado River channels. South of El Mayor the Highway follows the trace of the frontal fault of the Sierra Mayor. At the south end of the Sierra Mayor the Highway crosses the supertidal flats of the upper Gulf of California. This area is the entrance for the gulf waters into Laguna Salada.

***Sierra Pinta to San Felipe** - At the south end of the supertidal flat the Highway enters the highly faulted and very rugged Tertiary volcanic area of the Sierra Pinta. The low steep hills to the left of the Highway at the north end of the Sierra Pinta are calcareous Paleozoic sedimentary rocks. The dune field was blown from Laguna Salada. The Highway will then travel through several low passes in the volcanic rocks passing remnants of uplifted and dissected alluvial fans.*

South of the Sierra Pinta the Highway undulates along the lower edges of uplifted alluvial fans to the junction of the Ensenada Highway. The supertidal salt flats of the Salinas de Omotepec dominate the view to the east. Nearer the junction the high rugged hills on the right are the granitic and metamorphic rocks of the south end of the Sierra Pinta. The rounded volcanic hill on the left is El Chinero. South of El Chinero Junction the highway continues to traverse the uplifted Alluvial Fans along the edge of the Salinas de Omotepec to San Felipe.

0 This log of the Highway starts at the Junction of Mexico 2 and 5 at Kilometer 0.

SAN FELIPE DESERT: The San Felipe Desert is one of four sub-deserts of the Desert Phytogeographic Region. Generally, the sparsely vegetated, extremely arid, San Felipe Desert of northeastern Baja extends south from the International Border along the eastern escarpment of the two major peninsular ranges, the Sierra de Juarez and the Sierra San Pedro Martir; to Bahia de los Angeles. This desert is extremely arid because it is in the rain shadow of the peninsular ranges. The two dominant plants of this region are Creosote Bush and Bursage which comprise the Creosote Bush Scrub plant community (*See* 14:108).

2 As the highway crosses Rio Nuevo (the New River in the U.S.), notice the spillway gate and pond on the south side of the highway. Rio Nuevo flows north into the Salton Sea on the U.S. side of the International Border.

5.8 The highway crosses the meandering Rio Nuevo for a second time.

10 The highway heads to the south through the irrigated agricultural fields. The irrigation water arrives in the valley from the Colorado River via an extensive canal system from Presa Morales located near the town of Algodones.

The range of mountains to the right is the Sierra Cucapa. To the far right the higher ranges of the Sierra Juarez are visible. The low hill ahead to the left is Volcan Cerro Prieto. It is a Quaternary volcanic cone described in detail at Km. 16.6.

Nopal or Mission cactus (*Opuntia fincus indica*) is grown in many of the yards of the houses on the side of the highway.

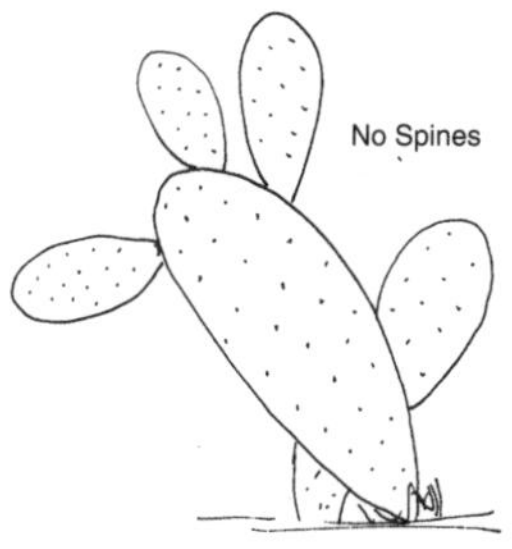

THE NOPAL CACTUS (*Opuntia*) belongs to the family Cactaceae and looks like beavertail cactus. Young Nopal leaves are larger and do not develop spines right away. They have little knobs which are not sharp. Nopales leaves are harvested young, shredded, and used to add fiber and bulk to the diet. They are also cooked like green beans, fire roasted, or pickled with chilies.

16.6 This canal marks the trace of the Cerro Prieto Fault Zone. Areas along fault zones in Baja including this one have been explored for geothermal energy (*See* 5:32).

The prominent 250 meters high dark peak to the left front at 10 o'clock is Volcan Cerro Prieto, a Quaternary volcanic cone with uplifted Tertiary sedimentary rocks on its sides. The Mexican government generates thousands of kilowatts of power (620 MW at this time) from geothermal wells located southeast of Cerro Prieto. These geothermal power plant facilities are the second largest in North America exceeded only by the "Geysers" geothermal facilities located north of San Francisco, California. There are hot mineral springs and mud volcanos in the area near a swampy area called Laguna Volcano.

21.2 Streams from the Mexicali Valley area are eroding into the alluvial fan to the right, establishing a new gradient on the fans at the new base level about 3 to 5 meters below the former base level. This has created a series of sharp arroyos and abandoned fan terraces in the sediments. To the south these dissected fans are quite obvious as the highway crosses the numerous alluvial fans of the Sierra Cucapa.

24 To the east the vegetation of San Felipe Desert is uniformly low and sparse. The flora of this region is predominantly Acacia, Smoke Trees, Bursage, and Ocotillo. The Acacia and Smoke trees are heavily parasitized by the evergreen mistletoe (*See* 16:15.5). On the right nearer to the Sierra Cucapa, the vegetation becomes taller.

The extensive quarry of Villegas, on the right, is developed in granitic alluvial sands.

25 The highway approaches and skirts the Sierra Cucapa (named for a local Indian tribe).

The Sierra Cucapa were uplifted along several fault zones which bound and diagonally cut across the range. In the central part the range is largely tonalite and granodiorite.

26 The plumes of steam in the distance to the left are produced by the geothermal power plants located near Volcan Cerro Prieto.

37 El Tave junction marks the intersection of Mexico 5 south and BC 18 east. South of this junction, the massif of the Sierra Cucapa is largely composed of granodiorite, prebatholithic schists, and calcareous metasedimentary rocks.

42 The Sierra Cucapa has declined into a low range of foothills. To the right front is a low pass which separates the Sierra Cucapa from the Sierra el Mayor which is the main mountain mass ahead.

49 At El Medanito the highway curves southeast along the base of the Sierra el Mayor. A cemetery is located in a series of stabilized dunes at the base of the Sierra el Mayor on the right.

49.5 The metamorphic rocks of the Sierra el Mayor are very close to the highway.

To the east there is a view of the Rio Hardy. The sloughs and banks of Rio Hardy provide excellent birdwatching.

50 The hills on the right have a series of smooth sloping small terraces developed on them. Each level represents a different episode of uplift of the Sierra el Mayor.

52 Campo Sonora is located to the southeast on Rio el Mayor, a stream tributary that flows into the larger Rio Hardy near Las Cabanas.

53 There is an excellent view ahead of the uplifted dissected terrace surfaces.

53.5 As the highway skirts the Sierra el Mayor, it passes an abandoned backwater slough of the Rio Hardy. The river has changed its course many times in the past and may even return to revive this abandoned slough at some future date.

55.5 The small settlement of Campo Rio Mayor Solano (La Carpa) is located on one of the major side channels of Rio Hardy.

56 The vegetation of this area is typical of slough vegetation and is dominated by masses of flat narrow sword-shaped monocot culms (stems).

Tamarisk, Broom Baccharis, Acacia, and Palo Verde are growing on the slopes of the foothills of the Sierra Mayor.

66 Because of the high salinity of the soils in this area only a few halophyte species are able to grow. Consequently, the vegetation to the left

(east) is dominated by almost pure stands of halophytic Creosote Bush and Bursage sparsely interspersed with Tamarisk.

69 As the highway heads south, it travels across a broad flat salt pan.

More granitic rocks of the Sierra el Mayor are exposed along the highway. The range directly ahead is the Sierra las Pintas. To the right of the highway across the salt pan, the Sierra las Tinajas, a lower range of mountains, are also visible. Behind them the higher range farther to the west is the flat crest of the Sierra Juarez.

73 The highway crosses the supertidal flats of the Gulf of California on a levee. Actually this area is the southeastern end of Laguna Salada, a basin located to the northwest (*See* 13:27.5).

80 A series of active and partially stabilized sand dunes are located along both sides of the highway. Creosote Bush and Bursage are growing on the dunes. In this area, on hot days, mirages are commonly seen on the salt pans around the Sierra las Pintas.

MIRAGES are often seen on the dry barren salt pans, supertidal flats, and playas of Baja. Mirages are optical phenomena that create the illusion of water. They are often complete with inverted reflections of distant objects identical to real bodies of standing water. They result from the distortion of light by alternate layers of hot and cool air. Mirages are also known as "fata morgana" because they are attributed to the witchcraft of Morgan le Fay, the sorceress sister and enemy of the sixth century British hero, King Arthur.

83 The Sierra las Pintas are the higher mountains to the right ahead. They are composed of a thick sequence of Tertiary volcanics. The hills to the left ahead are composed of upper Paleozoic meta-sedimentary rocks which contain Crinoid fossils. These were the first Paleozoic sedimentary rocks to be documented in Baja California.

87 On the left, a dune field has developed on the foothills of the Sierra las Pintas. They originated as winds blowing southwest across Laguna Salada picked up sand and deposited it on the eastern slopes of the range.

PALEOZOIC ROCKS WITH SAND DUNES

97.3 The highway travels along the eastern edge of the Sierra las Pintas for many kilometers. The broad white flat area to the left is part of the Salinas de Omtepec, a supertidal flat located at the northern end of the Gulf of California. Supertidal flats are inundated during very high tides.

105 The small settlement of La Ventana is located on the right side of the highway.

108 This is an excellent place to stop to take a walk and observe the flora of the Creosote Bush Scrub, the plant community characteristic of the San Felipe Desert area of the Desert Phytogeographic Region. The vegetation, which grows on the eastern foothills of the Sierra las Pintas consists of Creosote Bush, Bursage, Brittlebush, Candelabra Cactus, Mesquite, Spurge, Cat's-Claw Acacia, Smoke Tree, Desert Thorn, and Palo Verde.

THE CREOSOTE BUSH SCRUB PLANT COMMUNITY OF the SAN FELIPE DESERT: The highway from Mexicali to San Felipe passes through the Creosote Bush scrub plant community of the San Felipe Desert area of Baja's Desert Phytogeographic Region. This community is found below 900 meters on slopes, alluvial fans, and valleys in the desert. The precipitation is low and mostly comes in the spring with a few summer showers. The pedocal soils which support this community are usually well drained, and the extremes of temperature are great (*See* 3:152.5). The community is so named because Creosote Bush (*Larrea divaricata*) is the largest most abundant plant. Along with Bursage

(*Franseria dumosa*) these two plants compose nearly 90% of the foliage cover.

Creosote Bush is a shrub which is commonly 1 to 2 meters high with very small resin-covered evergreen leaves.

Bursage is a low rounded ashy-gray-green shrub 2 to 6 decimeters high with grayish-white bark and stiff intertwining branches.

The two dominant shrubs of this region, Creosote Bush and Bursage, look as though they have been hand planted in straight lines. However, they are naturally spaced as an adaptation for utilizing water more efficiently. Every plant requires a certain amount of water. In wet soils plants grow closer and more randomly; in dry environments plants are spaced almost territorially and grow evenly and widely spaced to ensure that each plant receives its required amount of water and minerals. The drier the environment, the wider and more regular the spacing. This is easily seen by noticing how widely spaced and short the Creosote Bush is on the tops of the hills. On the slopes Creosote Bushes become more numerous, more closely spaced, and taller. Taller and more closely spaced Creosote Bushes grow at the bottoms of the hills where runoff collects. The tallest Creosote Bushes grow along the shoulders of the highway where they receive the runoff from the highway and are able to grow their best.

Another factor responsible for the linear arrangement of Bursage and Creosote Bush in straight lines is the majority of these two dominant plants grow along tiny drainages, known as rills, that are less than five centimeters deep. Even small rills in desert landscapes may alter vegetational distribution patterns and community composition.

A third phenomenon that is responsible for the wide spacing of the Creosote Bush is allelopathy. If seeds of the Creosote Bush or other plant species were allowed to grow under or between the Creosote Bush, they would compete for the meager soil water and minerals available in the extremely arid nutrient-poor desert soils of the San Felipe Desert. To prevent competition resinous leachates from the Creosote Bush's leaves and stems "poison" the soil, which prevents the growth of other plants. Despite the barren appearance, quite a variety of desert species grow here forming a regional assemblage of plant species known as the **Creosote Bush Scrub**. They are:

CREOSOTE BUSH SCRUB

PLANT	COMMON LOCATION
Bursage or Burroweed (*Franseria dumosa*)	Widespread-usually with Creosote Bush
Creosote Bush (*Larrea divaricata*)	Slopes, washes
Ocotillo (*Fouquieria splendens*)	Slopes, flats, washes
Cheesebush (*Hymenoclea salsola*)	Washes, highway edges
Cholla (*Opuntia sp.*)	On rocky slopes and disturbed soils
Desert Willow (*Opuntia sp.*)	Infrequently along watercourses

123 The Sierra Pintas are highly mineralized. The La Fortuna and Moctezuma mines are located in the hills to the right. At one time silver and gold ore was extracted from both mines.

125 The flat region ahead to the right is Llano Cerro el Chinero.

134 The volcanic hill to the left is Cerro el Chinero. Years ago some Chinese immigrants were dropped off at San Felipe by an unscrupulous sea captain and told that Los Estados Unidos were just to the north. They walked until they got to this hill. Reportedly this is as far as they got before most of them died, still a long way from the United States.

The amphitheater in the center of Cerro el Chinero contains volcanic cinders indicative of an eruptive center. At the base of Cerro el Chinero sand dunes that have blown against the hill have been dissected by streams which gives the impression of cone-shaped alluvial fans. However, the wide part, not the narrow apex, of the fan is at the streams' head; the "fans" are upside down because they are really dunes. The dunes on the western flank of Cerro el Chinero are sparsely vegetated with small tufts of grass, scattered Brittlebush, and Skeleton Weed. Cerro el Chinero, itself, is covered by relatively dense stands of Ocotillo.

The flats are covered by coarse materials that have formed desert pavement.

DESERT PAVEMENT: Deflation (from the Latin meaning "to blow away") is an erosive process by wind that blows the unconsolidated sand and dust particles from a deposit and leaves the larger pebble or cobble-sized particles. Aided by alternating wetting and drying of the ground, the pebbles slowly rotate, settle, and concentrate into a flat armor-like mosaic called desert pavement. In some regions of Baja, the mosaic is made of several different kinds of rock, and patches occur where every stone is a different color. The pebbles which form the desert pavement on the lower slopes of Cerro el Chinero are a uniform brownish color. The even surface of the mosaic floor no longer offers any elevations that the wind can attack, further deflation ceases, and the erosion process ends.

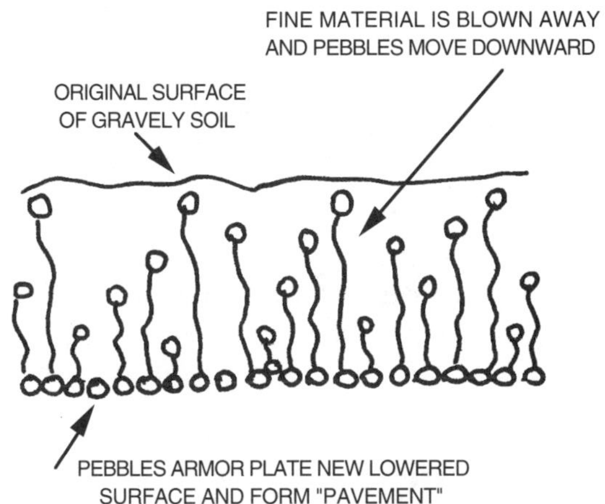

Do not be fooled by the hard look of desert pavement! It often superficially covers areas of soft deep sand, and a car driven onto desert pavement may sink up to its axles!

135 As the highway passes Cerro el Chinero, the high ridge to the right in the distance is the crest of Sierra San Pedro Martir and its high peak, Picacho del Diablo (Elev. 10,126') located in Baja's "Parque Nacional San Pedro Martir Constitucion, 1857".

136 Dense Cheesebush and large Brittlebush are growing along the highway. If you crush the leaves of Cheesebush between your fingers they produce a cheese-like odor.

Red-tailed Hawks (*Buteo jamaicensis*) are often seen here.

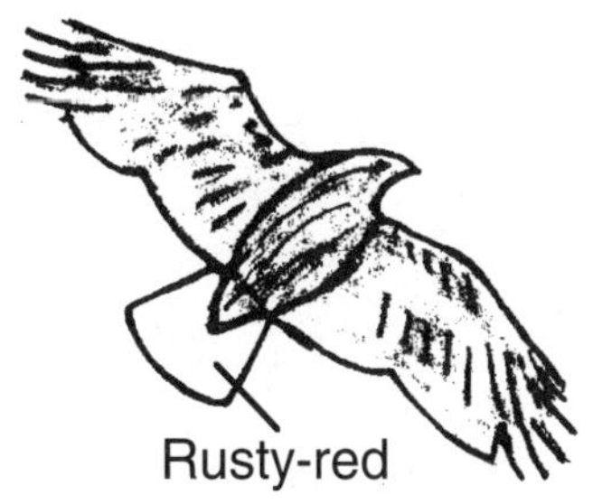

PREDATORY HAWKS OF THE DESERT: The **Red-Tailed Hawk or Aguililla Parda** is a well-known buteo commonly seen perched on top of telephone poles or fence posts along the highway and where there are open hunting grounds. Usually hunting alone, this hawk may sit motionless for hours, then suddenly emit a high scream, and swoop down to prey on rabbits, rodents, and other ground dwelling prey. Two field characteristics which make red-tails easily identifiable are their uniformly colored red tail and their dark belly band.

140 Junction of Highway 3 to Ensenada (195 Kilometers) and Tijuana (300 Kilometers). If you are heading to Ensenada turn to the right (west) and begin to cross the peninsula. (See log 12).

157 This is the first view of the Gulf of California and the extremely wide Salinas de Omtepec.

161 The shoreline of the Gulf of California approaches the highway from the Colorado River Delta area and the gulf island, Isla Montague. The slope that the highway is on goes down to the flat area of the salt pans of Salinas de Omtepec.

The Salinas de Omtepec are laced with little sloughs that flood during super high tides. They are supertidal flats which continue along the coast to the head of The Gulf of California. They are related to the very high tidal range in the gulf and the emptying of the Rio Colorado into the upper Gulf of California.

165 The brown hills to the right are desert varnished tonalite. The vegetation is similar to the flora or the Anza-Borrego Desert. The predominant plants are Ocotillo, Garambullo, abundant Mistletoe, Cat's-Claw Acacia, Mesquite, and Purple Bush. Opportunistic types of vegetation such as Cheesebush are growing along the side of the highway. It is almost the only green plant in this entire area.

THE AVIFAUNA OF THE SAN FELIPE DESERT: The bird species of the Gulf coast of Northeastern Baja are the same as those found in the Californian Region, the desert regions of mainland Mexico, and the Southwest U.S. The following sixteen species are typically seen along the northeastern peninsular Gulf shoreline and in the Creosote Bush plant community of San Felipe Desert between the International Border and San Felipe:

BIRD NAME	LIKELY LOCATION
Amer. White Pelican	Fly low over coastal waters
American Kestrel	On telephone wires
Anna's Hummingbird	Near red or yellow flowers
Bendire's Thrasher	On the ground in the brush
Cactus Wren	Flying or perching on cacti
Calif. Brown Pelican	Flying over coastal waters
California quail	On the ground in coveys
Coasta's Hummingbird	Near red or yellow flowers
Gila Woodpecker	On poles or in large plants
Gilded Woodpecker	On poles or in large plants
Greater Roadrunner	crossing the highway
Ladder-back Woodpecker	On poles or in large plants
LeConte's Thrasher	On the ground in the brush
Red-Tailed Hawk	perching on poles or soaring
Turkey Vultures	feeding on carrion or soaring in circles overhead
Western Meadowlark	Fences bordering meadows

167 To the left at the highway is Cerro el Moreno, a Miocene volcanic plug. Nearby to the east there is a small inlet on the coast, Estero Primero, which is in the southern region of Salinas de Omtepec.

170 The high hill to the right in the near distance is Cerro el Colorado which is composed of Tertiary volcanics. The lower hills in front are all tonalite. The low darker-gray hills to the southwest are composed of prebatholithic slates and schists.

175 In the distance to the west are the high peaks of the Sierra San Pedro Martir and the eastern escarpment of the Peninsular Range Batholith. The massif of the Sierra San Pedro Martir is the highest range on the peninsula of Baja. The highest peak of the range is Picacho del Diablo (Devil's Peak; 3115 m., 10,126'). It is located to the right at approximately 3 o'clock. The mountains in the foreground are the Sierra San Felipe.

177 For the next several kilometers the highway passes semi-stabilized dune fields of sand blown from the sandy beaches of the Gulf.

There is a view of the Gulf of California, and if the weather is clear, Isla Consag, a steep sided volcanic peak may be visible in the middle of the Gulf 30 kms. offshore. This isolated volcanic peak, was named for Padre Fernando Consag, an early explorer the Gulf of California (1746).

179 The road to the right leads northwest to Colonia Morero and El Saltio. This sandy road crosses Valle Santa Clara and goes north along Laguna Diablo to intersect the main Mexico Highway 3 east of Valle Trinidad (See 12:175.5).

182 To the left is a "living" fence made from Ocotillo cuttings. Some of the posts have rooted by adventitious roots, so the plant has started to grow again. This is similar to the living Datillo fences seen in other parts of the peninsula (*See* 4:82.5).

187 The highway passes through a low range of hills of faulted prebatholithic carbonate rocks. They are covered by desert-varnish which gives their surface a very dark appearance (See 14:134)

189 As the highway enters San Felipe, Cerro el Machorro and the point of Punta el Machorro which are composed of tonalite and granodiorite are to the left. The rocks to the left at the monument circle are prebatholithic carbonate rocks.

Log 15 - Ensenada to Tecate [114 kms = 71 miles]

The Highway travels up a gentle alluvial valley flanked by steep hills of the marine Rosario Formation, then climbs a steep grade into the rugged metavolcanic hills. After passing through the metavolcanic hills, the Highway travels through a rolling bouldery area of tonalite on the south side of the Valle Guadalupe eventually travelling on the alluviated floor of the valley with steep metavolcanic and granitic hills in the distance on both sides. The Highway then travels through more granitic rocks and crosses a fault into an area of rugged hills in slatey metamorphic rocks with scattered granitic rocks. The Highway enters steep bouldery tonalite hills near El Testerazo and climbs a grade through the conglomerates of the Las Palmas Gravels. On top of the grade the Highway crosses the flat surface of the conglomerates with views to the north and east of the flat to undulating erosion surface with granitic peaks, and views to the west and south of the rugged metavolcanic foothills.

The Highway then descends a grade through the conglomerates to the alluviated Las Palmas Valley. North of the valley the Highway climbs through steep tonalite hills and then through the bouldery granodiorite. The Highway follows a spectacular contact between a dark, smooth, gabbro hill on the right and a light, bouldery, granodiorite hill on the left then travels through the rugged granodiorite hills past several gabbro hills to descend steeply through tonalite into Tecate.

104.5 This log begins at the junction of Mexico 1 (*See* 1:101.3) and Mexico 3 and ends at Tecate. (*See* 13:131.5) The highway initially goes inland in the alluviated valley of Rio San Antonio. It is bordered by hills of Cretaceous rocks on both sides.

101 The highway begins to climb through prebatholithic metavolcanic rocks.

98.7 The roadcut at the top of the grade exposes a massive purple andesitic breccia, a common rock type in the prebatholithic metavolcanic rocks. The highway continues to pass through the metavolcanic rocks for the next 2 kilometers.

96 The highway enters the southwestern end of a valley in the Oak woodland. The highway passes through numerous Oak woodlands intermixed with chaparral areas all the way to Tecate.

92 The road to the left to Alegre (El Tigre) connects with the Free Highway between Tijuana and Ensenada. The boulder-covered slopes of this region are tonalite and the smoother slopes are generally metavolcanic rocks and slates.

88 The flat hill to the left is a remnant of a terrace level from an older alluvial fan.

83 The main part of Valle Guadalupe is to the left.

The hills on the far side are prebatholithic metavolcanic rocks and tonalite.

77.4 The highway crosses Rio Guadalupe which originates in the Sierra Juarez near Laguna Hansen and meanders to the west through Ojos Negros, Valle Guadalupe, and to the sea at La Mision. Most of the water flow occurs underground, as ground water, leaving the streambed surface dry. The vegetation in the streambed consists of Tamarisk, Broom Baccharis, and Willow.

76.8 This turnoff to the left leads to Colonia Guadalupe, the site of a Russian agricultural colony established in 1905. The Dominican Mision Nuestra Senora de Guadalupe was also established here in 1834. It was the last mission to be established in Baja California.

The highway follows the west side of Guadalupe Valley at the base of the tonalite and granodiorite hills.

GUADALUPE VALLEY AND TONALITE HILLS

71.5 The highway follows a fault line valley in tonalite.

70 For the next several kilometers many small Riparian Woodland areas are seen along the right side of the highway. The dominant vegetation along the stream banks is Coastal Live Oaks, Scrub Oak, Sycamores, and Willows. Acacia, Broom Baccharis, Chamise, Toyon with Dodder, and Wild Buckwheat are the plants which grow on the low surrounding hills.

65 Prebatholithic slates, phyllites, and argillites outcrop along the highway. These outcrops are probably Paleozoic in age. To the east there are outcrops of Ordovician quartzite, chert, argillite, and marble (Lothringer, 1984)

64 The Riparian vegetation has been temporarily left behind prior to entering the valley of Ignacio Zaragoza where the highway passes through a Coastal Sage Scrub community which consists of Rabbitbrush, Scrub Oak, and several species of composites.

Numerous granitic dikes to the east cut the metamorphic rocks. The well-known ring dike of El Pinal (Duffield, 1969) is a few miles to the east.

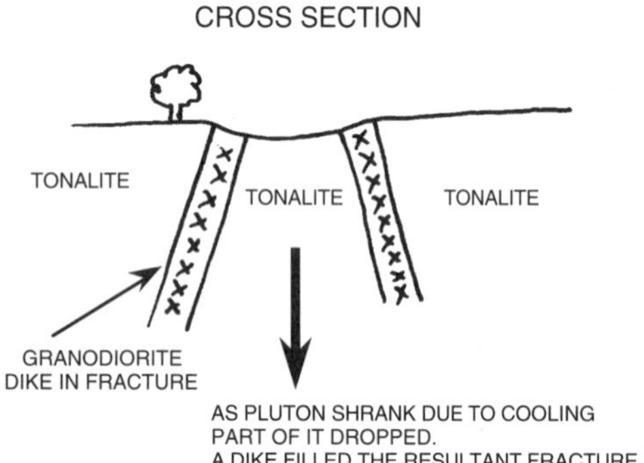

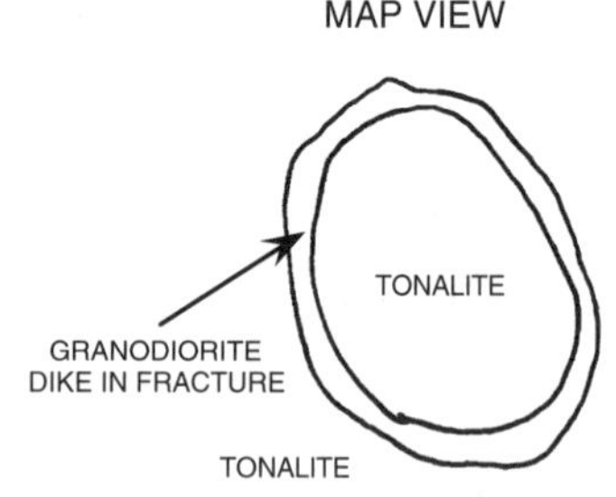

Ring Dike: Dikes form as molten material fills a crack in the rocks. A small pluton of granitic rock was intruded into the El Pinal area. As this pluton cooled, it cracked parallel to its circular outline. The resultant crack was filled by the dike material and left a "ring dike" which is prominent on the Baja geologic map.

The large green trees along both sides of the highway are Tamarisk, Cat's Claw Acacia, Cottonwoods, and Willows.

The major northwest-southeast-trending Vallecitos Fault Zone crosses the highway near the south end of the valley. This fault zone may continue through to the Tijuana River Valley.

63 The roadcuts for the next 5 kms. expose dark-colored slatey to phyllitic prebatholithic rocks.

PREBATHOLITHIC SLATES AND PHYLLITES

62 A beautiful parklike oak woodland extending along the drier banks of a Riparian woodland is now visible to the left side of the highway. The large silver-barked trees are Sycamores; the dark-green trees are Oaks..

62.5 The highway passes through roadcuts with good exposures of schist and gneiss cut by dikes.

60 The highway has entered the broad alluviated Vallecitos which is ringed by metavolcanic hills with a few scattered patches of tonalite. Dikes cut some of the hills.

55.4 The lower part of the small valley ahead is vegetated with Riparian cottonwood trees that are heavily parasitized by mistletoe.

TONALITE BOULDER

52 The hill of conglomerate ahead has been uplifted between two branches of the Falla de Calabasas. The highway has been built in tonalite. Numerous small shear zones cut the rocks for the next kilometer.

The valley at Kilometer 51.2 is a fault-line valley. The highway runs parallel to the fault zone at the Kilometer 51 mark.

50.5 The highway crosses Rio las Calabazas. This fault-line valley parallels the one at Km 52.

49 El Testerazo. The hills ahead are composed of Eocene conglomerates which were deposited 45 million years ago by a major river which flowed across the range from Sonora to the Tijuana area.

48 The high hills to the left are metavolcanics; to the right a tonalite body is visible.

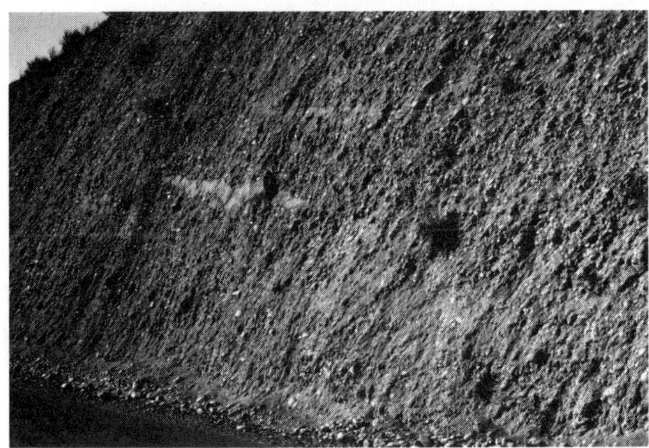

CONGLOMERATES FROM AN EOCENE RIVER

47 As the highway climbs a grade, it passes through Eocene conglomerate outcrops.

45 The conglomerates are more firmly cemented in this area. The red color of the cuts is due to slope wash; that is, the red iron oxides from upper sedimentary rocks have been leached out and have run down over the face of the lower rocks coating them with a reddish wash.

43 The vegetation of this area is the Coastal Sage Scrub plant community (*See* 1:14.8).

EDGE OF EOCENE RIVER CHANNEL

40.5 To the right is a good place to view the conglomerates and erosion surfaces. The conglomerates which comprise these intervening hills are part of a string of conglomerate exposures which stretch to the crest of the Peninsular Ranges to the east. They contain a small percentage of rhyolite clasts similar to those of the Poway Conglomerate of the San Diego area (Minch, 1972, 1984). The conglomerates rest upon an Eocene or older erosion surface. The erosion surface can be seen to the east (*See* 13:109).

Cerro La Libertad to the northeast was a monadnock on this surface. Monadnocks are hills representing scattered erosional remnants on a low-relief erosional surface (peneplain). West of this area the surface has been largely destroyed by stream erosion.

The hills to the west and southwest are across a fault and part of the metavolcanic coastal range. Metavolcanic rocks do not generally develop erosion surfaces.

EROSION SURFACE FROM KILOMETER 40.5

38.9 Metavolcanic outcrops which formed part of the north bank of the Eocene channel are visible in this roadcut.

38.6 The broad valley to the right has been cut in tonalite.

AREA OF TONALITE OUTCROPS

37 The vista opens to the tonalite hills cut by several light-colored granitic dikes. The light-colored linear bouldery outcrops are erosional remnants of these granitic dikes. Two granodiorite peaks form the skyline: Cerro los Bateques is the lower and Cerro La Libertad is the higher peak.

35.8 The highway passes through decomposed pebbly mudstones and sandstones which underlie the conglomerates of this region. These conglomerates represent a pre-Eocene channel similar to the Cretaceous Lusardi conglomerates in the San Diego area. They were deposited by short streams as the batholith was uplifted and eroded during Mesozoic subduction. These older conglomerates are only exposed on this side of the Eocene valley.

A CRETACEOUS VALLEY IS FILLED WITH COARSE GRAVELS

AN EOCENE RIVER CUT A VALLEY WHICH OVERLAPPED THE OLDER CRETACEOUS CHANNEL

THIS VALLEY IS THEN FILLED WITH EXOTIC GRAVELS

PRESENT EROSION HAS ERODED THE SURROUNDING ROCKS AND REMOVED THE SIDE OF THE VALLEY AND LEFT A CONGLOMERATE RIDGE

32 The Falla de Calabasas that the road crossed near El Testerazo is seen near the break in the slope ahead. It passes through the notch between the two very high hills, and across the valley to the right of the zigzag in the Cerro Bola road.

31 For the next 3 kilometers the highway passes along the edge of Valle de Las Palmas. The

highway between here and Tecate passes through granitic rocks.

CRETACEOUS PEBBLY MUDSTONES

30.5 The road to the left climbs Cerro Bola where Microondas Cerro Bola is located. The lower part of the road crosses a smooth gently sloping surface which is part of the old alluvial fans that developed in this area before it was uplifted. The upper part is in the metavolcanic rocks.

28 The highway crosses Arroyo Seco.

25 The highway passes through an extensive area of bouldery outcrops of granodiorite.

GABBRO VS GRANODIORITE OUTCROPS

16 Notice the contact, along the highway, between a granodiorite hill on the right and a gabbro hill on the left. The gabbro weathers easily and has no outcrops, while the granodiorite hill weathers more slowly and has bouldery outcrops. As the gabbro weathers, it forms clay soils. These clay soils protect the underlying gabbro from further weathering; this results in fresh gabbro relatively near the surface. However, the granodiorite weathers to a sandy soil which allows further deep weathering leaving the surface boulders on a weathered base.

15 The highway crests over a hill with a view of the erosion surface.

7 The high metavolcanic peak in the distance is Tecate Peak in the U.S.

3 The descent into Tecate begins as the road passes under the new Tijuana to Mexicalli Toll Road. The numerous brickyards in this area exploit the clay soils developed on isolated gabbro bodies.

0 The highway crosses railroad tracks at Tecate on Calle Ortiz Rubio. The main Tijuana to Mexicali highway, Avenida Juarez, is at the second stop sign. Tijuana is 30 miles to the west of Tecate. Those who wish to continue directly to the United States should turn left for one block (to signal) then turn right on Calle Lazaro Cardenas to the Tecate border crossing. This log joins the Tijuana-Mexicali log 13:131.5.

Cerveceria Tecate, one of the largest breweries in Mexico, is two blocks left at the first stop sign (on Avenida Hidalgo).

It is worth noting that while this area seems to be arid, the world's record for high intensity rainfall of 11" in 80 minutes was set in the small town of Campo just north east of Tecate. Campo was a rail camp when they were building the San Diego & Arizona Eastern Railroad.

Log 16 - L.A. Junction to Bahia de Los Angeles [66 kms = 40 miles]

The Highway gently climbs eastward on the flat surface of the Miocene fluvial sedimentary rocks toward a low rolling series of metamorphic hills. It then drops into an alluvial wash and climbs out to pass south of rugged rhyolite hills then between a series of rugged metamorphic hills in a broad sloping alluvial valley. The hills and mesas to the north are capped by rhyolite and basalt. The Highway then steeply descends a pass in metamorphic rocks to skirt the south side of the dry lake of Laguna Amarga along a series of rugged metasedimentary hills.

The opposite sides of the playa are a steep, rugged series of faulted, metamorphic and granitic rocks capped by mesas of basalt tuffs, and rhyolite. After crossing the southern part of the lake the Highway again descends a steep canyon in rugged hills of metasedimentary rocks to the alluvial fans on the shore of Bahia de Los Angeles. Here the Highway turns south on the fans and parallels a fault scarp in the granitic and metasedimentary rocks.

0 About 10 kilometers north of Punta Prieta is the junction of the highway and a paved road that heads east toward Bahia de Los Angeles. For the first ten kilometers the highway gently climbs on the surface of the Miocene fluvial sedimentary rocks that mantle this relatively flat area. The mesa to the right consists of volcanic and fluvial sedimentary rocks capped by dark basalt. As the highway continues eastward it approaches Mesa La Pinta, a granitic and metamorphic hill.

The vegetation consists of species typical of the Vizcaino Desert flora (*See* 3:69). Cirio, Ocotillo, Elephant Tree, and Cardon are the dominant taller plants. Datillo, Yucca, Cholla, Garambullo, and numerous species of desert annuals form the low "understory". An occasional Palo Verde or Mesquite is visible. The Mesquite in this area is heavily parasitized by Mistletoe.

5.5 Hills of rounded granitic rock and metamorphic rocks can be seen about one half kilometer to the left of the highway.

10 The highway enters an area of low hills with metamorphic and granitic rocks exposed on both sides of the highway. Many of the granitic rocks are spheroidally weathered.

14.5 Here the highway dips into an alluvial wash primarily vegetated by Datillo and Cardon.

15.5 As the highway quickly climbs back onto the surface of the Miocene fluvial sedimentary rocks notice the epiphytic, pineapple relative Ball Moss abundantly covering the Cirio, Desert Aster, Skeleton Weed, Atriplex, Mesquite, Desert Hollyhock, Coyote Melon, and Indigo Bush, Agave, Elephant Tree, and Ocotillo grow in this area.

EPIPHYTES AND PARASITES: From south of El Rosario along the Pacific coast to the Cape, trees, shrubs, and cacti like Acacia, Palo Adan, Cirio, Ironwood, Lomboy, Mesquite, Broom Baccharis, Cardon, and Pitaya Dulce, are often covered with the moisture loving, epiphytic Ball Moss (*Tillandsia recurva*), a member of the pineapple family (*See* 3:74.5). Epiphytic Ball Moss is not harmful to the host (primarily Cirio here) plant since it lives in a commensalistic relationship depending on the host plant for support only. However, a close look at some of the same trees and shrubs listed above will reveal several harmful parasites living in a destructive symbiosis with them. The two most common parasitic organisms (symbionts) of this area are the evergreen Desert Mistletoe (*Phrygillanthus sonorae*) and "witch's hair" (*Cuscata veatchii*).

"Witch's hair", a plant lacking chlorophyll, attaches its yellow or orange, spaghetti-like stems to its host (Elephant Trees here) by a modified stem called a haustoria. Since it cannot make its own food it uses its haustorial stem to "steal" sugar produced photosynthetically by its host.

The second parasite common to this region is the woody, evergreen mistletoe seen hanging in large, shapeless masses, in the branches of many plants in Baja (primarily Mesquite here). They look like clumps of vegetation that are the wrong shape and wrong color, not matching the foliage of the host plant. Like "Witches Hair", Mistletoe utilizes haustorial stems to siphon sugar from their host. If the mistletoe is thick enough it will kill its host and itself at the same time. However, strong, healthy host plants can withstand the parasite attacks.

Along the Pacific coast, the moisture that sustains both the epiphytes and parasites of this desert, comes from fogs that form at night and lie during the early morning hours in misty layers among the hills until the sun "bakes" it off. Along the Pacific coast, from the Cape Region northward to San Francisquito, areas similar to this are known as coastal fog deserts. With only a few exceptions, these plants are unknown on the drier Gulf side of the peninsula where fog does not normally occur.

17 The pinkish colored flat-lying hills of the Mesa Tinajas Coloradas, off to the left of the highway, are Miocene extrusive rhyolites overlying the Miocene fluvial sedimentary rocks.

18 For about one half kilometer the highway skirts a dark colored, gabbro hill to the left of the highway. Very few gabbro hills are accessibly exposed along Baja's highways. Gabbros are dark colored basic, intrusive igneous rocks composed principally of labradorite or bytownite and augite. It is the approximate intrusive equivalent of the extrusive basalts.

23 The vista opens up ahead, as the highway descends a broad alluvial plain on the Pacific coast drainage, into the Valle Agua Amarga.

Along the right side of the highway is an Elephant Tree forest. Although the Elephant Trees look "dead" most of the year, during the rainy season they put out leaves and look lush and green. When they flower they look as though they are capped by a pinkish haze. All of the well-drained drier slopes in this area are covered with beautiful, thick stands of them. Cirios and Cardons vegetate the lower, flatter elevations.

30 As the highway crests a low pass it enters the Gulf drainage.

The hill to the left is a mixture of rhyolite and andesite overlying the Miocene fluvial sedimentary rocks which the highway is passing through.

31 At the crest of the pass, the vegetation consists of Datillo, Garambullo, Agave, Greasewood, Cirios and Cardon. The dominant tall plant here is the thick-trunked Elephant Tree.

32.5 The hills directly ahead are prebatholithic metasedimentary rocks with resistant light colored layers. The fine-grained, white bedded layers along the highway are Quaternary lacustrine sedimentary rocks which may represent dissected remnants of a tilted ancient lake bed.

33 The vista opens to Playa Agua Amarga, the dry lake bed at the bottom of the Valle Agua Amarga. Agua Amarga, bitter water, refers to the highly saline, unpotable waters found in the valley. The Sierra el Toro, located on the far side of the Valle Agua Amarga, consists of a mixture of granitic, metamorphic, sedimentary and volcanic rocks.

PLAYA LAKE: Playas are the flat, dry, barren area of an undrained desert basin underlain by clay, silt, and commonly soluble salts. It may be covered by a shallow, intermittent (ephemeral) lake during the wet season that dries up, by evaporation, during the hotter summer months. When the standing water of the ephemeral lake evaporates, it leaves behind a playa covered by salt deposits.

Valle Agua Amarga was a deeper valley, but valleys get filled up over geologic time. The deep valley fill is composed of the silt, salt, and sand which was originally weathered, eroded, or dissolved from the bedrock of the surrounding highlands and deposited in the valley, a process that is still occurring.

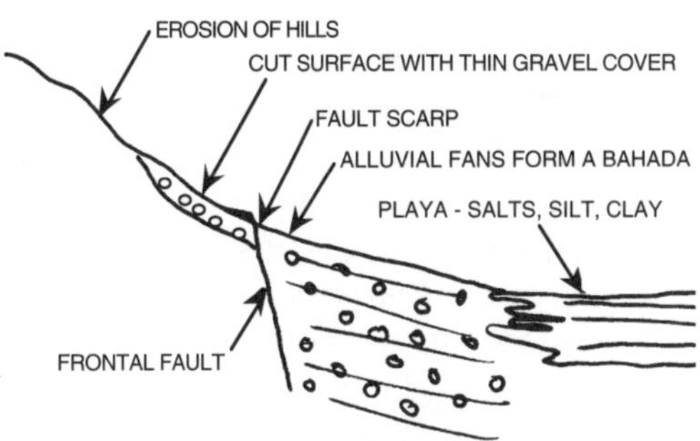

The highway skirts the Playa Agua Amarga salt pan. The barren, highly saline, white central portion of the playa is clearly visible and is totally devoid of life. However, the less saline soils along the edge of the playa are vegetated by a few hearty salt tolerant halophytes (salt lovers).

SALT PAN: Salt ($CaCl_3$, KCl, NaCl) and mud make up most of the flat floor of the Valle Agua Amarga. However, several other salts are mixed in with the Halite and mud in variable proportions. Originally rainwater dissolved all of the various types of salts from the bedrock of the surrounding mountains. Runoff and groundwater seepage have slowly moved the salts downward toward the lower parts of the valley floor resulting in the formation of the playa. Different salts move at different rates and so end up in different localities depending on their solubility in water.

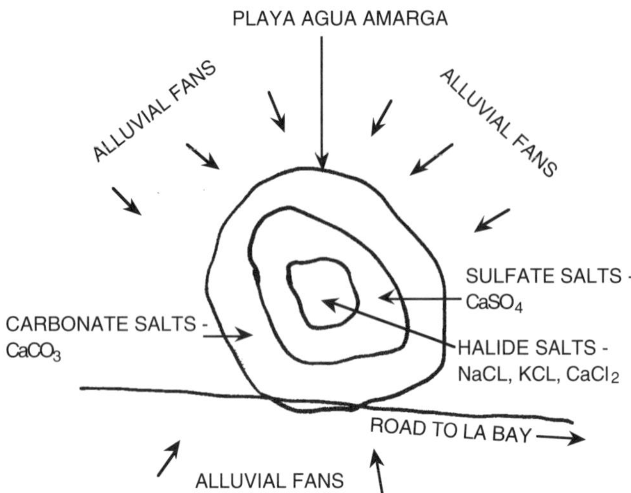

The salt in the soils of the alluvial fans is predominated by the least soluble carbonate salt Calcite ($CaCO_3$). Calcite crystallizes from groundwater as a natural cement and is commonly visible as white seams of caliche (*See* 3:152.5).

Moving down the fan, to the edge of the playa, the slightly more soluble sulfate Gypsum (hydrous $CaSO_4$) becomes abundant both as caliche-like veins and as gypsum crystals that have formed by efflorescence. Efflorescence is the process by which crystals grow, on surfaces, due to the evaporation of salt-laden water.

The central, parts of the Playa Agua Amarga are mostly Halite (NaCl) and other chlorides ($CaCl_3$, KCl) since they are the most soluble salts. Thus, there appears to be a crudely concentric zonation around the salt pan, reflecting increasing solubility, from carbonates at the top of the slopes to sulfates and finally to the highly soluble chlorides concentrated on the barren, lifeless floor of the salt pan.

35.5 As the highway begins to skirt the barren edge of Playa Agua Amarga the rocks along the right side of the highway are prebatholithic metasedimentary rocks and schists.

WHY IS PLAYA AGUA AMARGA NEARLY DEVOID OF VEGETATION?: Plants are extremely sensitive to changes in water quantity and quality. A sudden decrease in plant height and/or a change in the species composition of an area can give clues as to the quantity of water available and the quality of that water. Plants needing permanent water supplies, such as willows, alders, ferns, and rushes, are called phreatophytes (well plants). Plants needing little water to survive, such as Baja's Mesquite, Greasewood, Palo Verde, and various cacti, are called xerophytes (dry plants). No matter how dry the soil, there will be some plants that have adapted to the low level of water. However, few, if any, plants can tolerate more than 6% soil salinity. The high salinity of playa soils is the result of runoff, carrying dissolved salts that collect and concentrate in the low, central portions of playas.

In the desert and on playas like Agua Amarga, areas of ground are often devoid of vegetational cover. This most likely indicates that the salinity of the soil is greater than 6%. There may even be standing water but with a salinity of greater than 6% a "physiological desert" exists to which no plant can adapt. Notice that a few plants are seen growing on low hummocks in the playa and along the margins of the playa. This indicates that the soil salinity there is less than 6%.

39 Notice the dramatic vegetation change that occurs as you look from the slopes down onto the playa salt pan. These changes reflect slope drainage and increasing soil salinity. The hills to the right are covered with Elephant Trees; the lower slopes are covered with Cardon, Jumping Cholla, and Ocotillo; the flats are densely vegetated with Creosote Bush, Teddy Bear Cholla, and Greasewood. The very Saline Salt flats are bare.

41 Dense stands of Smoke Trees are growing in the wash on both sides of the highway. Smoke Trees only grow in washes since their seeds must be scarified by the sands of the wash during wet periods and water is necessary to leach out germination inhibitors. (See 13:43.5).

41.3 The sign here indicates there is a rough, back way to Mision San Borja recommended for 4-wheel drive vehicles only. Because of better road conditions, the road from El Rosarito (4:52.5) is the preferred route. Agua Higuera spring and cattle ranch are located about 1 kilometer up the wash to the right. The road then continues on to Mision San Borja (*See* 4:52.5).

43.8 This side road also leads to Mision San Borja. Both this road and the one at kilometer 41.3 meet at Agua Higuera spring.

45 The vegetation of this area is typical of that commonly found growing on similar flat areas along this section of the highway. Ocotillo, Cardon, Mesquite, Datillo, two species of Agave, Jumping Cholla, Old Man Cactus, Cirio, and several species of annual composites are growing at the foot of the slopes. On the higher, drier parts of the slopes, Elephant Trees, Teddy Bear Cholla, Jumping Cholla, and Cirios grow in abundance. The vegetation growing along the pavement edge are Cheese Bush, Brittle Bush, Nightshade, and Desert Hollyhock.

48 The highway crosses Playa Agua Amarga.

48.5 The Elephant Tree on the left is nearly covered by parasitic "Witches Hair".

52 The alluvial fan slope descending toward the highway from the right is densely covered with Elephant Trees and the smaller, more delicate, green-trunked Palo Verde.

53 Along with the first view of the gulf, the 75 kilometer long bulk of Isla Angel de la Guardia is visible to the northeast. It forms the distant line of hills in the gap in the peninsular hills to the left. The highway descends along an alluviated stream course with prebatholithic meta-sedimentary rocks exposed along both sides.

There are a large number of Cirios and a few Cardons in this area. From this point the Cirio will gradually disappear leaving the other vegetation relatively unchanged.

55 A number of impoverished Elephant Trees and Ocotillo are growing on the hills in this area.

56 The vegetation of this wash includes species characteristic of "Wash Woodlands". It is predominated by Smoke Trees, Palo Verde, Ocotillo, Greasewood, Wax Plant, Ephedra, Cholla, a few large scattered Cardon and Cirio, and assorted species of low annual scrub.

56.7 The highway enters a canyon cut in the prebatholithic metasedimentary rocks. These metasedimentary rocks are intruded by numerous light colored dikes.

58 To the right, below the highway, there is a gorge in the volcanics which was cut as an old stream meander was deepened by stream rejuvenation and uplift. Originally the stream cut a surface on the metasedimentary rocks. As a result of uplift, the stream cut a gorge 3-5 meter deep leaving the former surface as a bedrock terrace on each side of the gorge

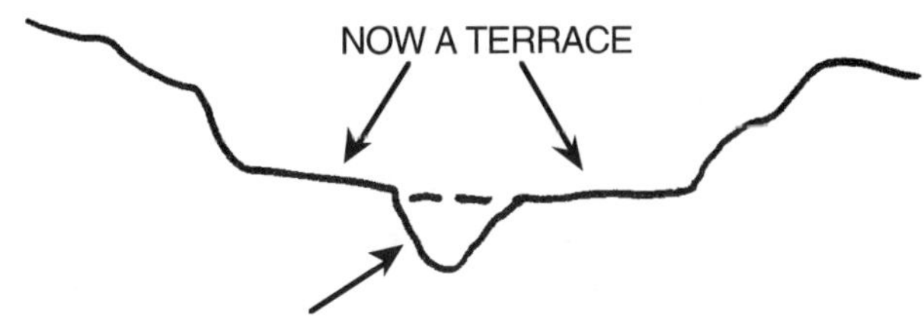

58.5 The view is of the small, closer islands in the Bahia de Los Angeles. Isla Angel de la Guardia dominates the center of the bay.

60.5 The highway drops down to an alluviated surface and turns to the south following the coastline to the main part of Bahia de Los Angeles.

62.3 The town of Bahia de Los Angeles comes into view with the sandy point and lighthouse of Punta Arena visible to the left and the south point of Punta La Herradura ("horseshoe"), which forms the outer point of the bay, in the background.

The dry slopes around this horseshoe shaped bay are sparsely vegetated by Ocotillo, Palo Verde, Creosote Bush and small Elephant Trees.

66 **BAHIA DE LOS ANGELES** is protected by the "midriff" islands producing a relatively calm bay loaded with corbina, yellowtail, sea bass, and other popular game fish.

WHERE TO NOW? To leave Bahia de Los Angeles either return the way you came or inquire locally about taking the 4 wheel dirt road leading southward through Las Flores, El Progresso, and El Arco that joins the main highway (Mexico 1) between Guerrero Negro and San Ignacio.

REFERENCES

Allison, E.C., 1974, The type Alisitos Formation (Cretaceous, Aptian-Albian) of Baja California and its bivalve fauna in Gastil, G., and Lillegraven, J., editors, Guidebook; the geology of peninsular California: Los Angeles, CA., American Assoc. of Petrol. Geologists, Pacific Section, p. 20-59.

Anderson, C.A., 1950, 1940 E.W. Scripps cruise to the Gulf of California, Part I: Geology of islands and neighboring land areas: Geological Society of America Memoir, 43, 53 p.

Ashby, J.R., and Minch, J.A., 1984, The upper Pliocene San Diego Formation and the occurrence of Carcharoden megalodon at La Joya, Tijuana, Baja California, Mexico *in* Minch, J.A., and Ashby, J.R., *eds*, Miocene and Cretaceous depositional environments, NW Baja Calif., Mexico: Ann. Mtg., Pac. Sec., Amer. Assoc. of Petrol. Geol. 54, p. 19-28.

Ashby, J.R., Ku, T.L., and Minch, J.A., 1987, Uranium-series Ages of Corals from the Upper Pleistocene Mulege Terrace, Baja California Sur, Mexico, Ciencias Marinas, v. 15, p. 139-141.

Ashby, J.R., Ku, T.L., and Minch, J.A., 1987, Uranium-series Ages of Corals from the Upper Pleistocene Mulege Terrace, Baja Calif. Sur, Mexico, Ciencias Marinas, v. 13, # 2.

Ashby, J.R., and Minch, J.A., 1987, Late Tertiary Sedimentation and Molluscan Paleoecology of the Mulege Embayment, Baja California Sur, Mexico, Geology, v. 15, p. 139-141.

Axelrod, D.I., 1978, The origin of coastal sage vegetation, Alta and Baja Calif.: Am. Jour. of Botany, v. 65(10), p. 117-131.

Beal, C.H., 1948, Reconnaissance of the geology and oil possibilities of Baja California, Mexico: Geol. Soc. of Am., Mem., 31, p. 138.

Borcherdt, R.D., 1975, Studies for Seismic Zonation of the San Francisco Bay Region, U.S. Geol. Sur. Prof. Paper, 941-A.

Case, T.J., and Cody, M.L., 1983, Island Biogeography in the Sea of Cortez, Univ. of Calif. Press, Berkley, 508 p.

Delattre, M.P., 1984, Permian Miogeoclinal Strata at El Volcan, Baja California, Mexico: *in* Frizzell, V.A., Jr., ed, Geology of the Baja Calif. Peninsula: Pac. Sec. SEPM, p. 23-30.

Duffield, W.A., 1968, Petrology and structure of the E1 Pinal Tonalite, Baja California, Mexico: Geological Society of America Bulletin, v. 79, p. 1351-1374.

Fife, D.L., Minch, J.A., and Crampton, P.L., 1967, A Late Jurassic Age for the Santiago Peak Volcanics, Geological Society of America Bull., v. 78, p. 299-304.

Fife, D.L., 1968, Geology of the Bahia Santa Rosalia Quadrangle, Baja California, Mexico [Master Thesis]: San Diego, CA, San Diego State College, 100 p.

Forman, J.A., Burke, W.H., Jr., Minch, J.A., and Yeats, R.S., 1971, Age of the Basement Rocks at Magdalena Bay, Baja California, Mexico, Sixty-seventh annual meeting, Cordilleran Section, Geol. Soc. of America, Riverside, Calif.

Gastil, G., Minch, J., and Phillips, R.P., 1983, The geology and ages of the islands in Case, T.J., and Cody, M.L., editors, Island biogeography in the Sea of Cortez: Berkeley, CA, University of California Press, p. 13-25.

Gastil, R.G., Phillips, R.P., and Allison, E.C., 1975, Reconnaissance geology of the State of Baja California: Geol. Society of America Memoir, 140, 170 p.

Gastil, R.G., Krummenacher, D., and Minch, J., 1979, The record of Cenozoic volcanism around the Gulf of California: Geological Society of America Bulletin, v. 90, p. 839-857.

Gastil, R.G., and Miller, R.H., 1984, Prebatholithic Paleogeography of Peninsula California and Adjacent Mexico in Frizzell, V.A., Jr., editor, Geology of the Baja California Peninsula: Pacific Section SEPM, p. 9-16.

Harris, M.E., 1991, Geology of Baja California: A Bibliography, San Diego State Univ. Library and Calif. Mines and Geol.

Hawkins, J.W., 1970, Petrology and possible tectonic significance of late Cenozoic volcanic rocks, southern California and Baja California: Geol. Soc. of Ameri. Bull., v. 81, p. 3323-3338. (*Also in* University California, Scripps Institution of Oceanography Cont., v. 40(2862), p. 1385-1400.

Heim, A., 1922, The Tertiary of southern Lower California: Geological Magazine, v. 59, p. 529-547.

Kilmer, F.H., 1963, Cretaceous and Cenozaic stratigraphy and paleontology, El Rosario area, Baja California, Mexico [Ph.D. dissertation]: Univ. of Calif., Berkeley , 216 p. 203.

Kilmer, F.H., 1965, A Miocene dugongid from Baja California, Mexico: Southern California Academy of Sciences Bulletin, v. 64, p. 57-74.

Knappe, R., Jr., 1974, The micropaleontology of a section of the Tepetate Formation, southern Baja California, and a paleobiogeographic comparison with equivalent foraminifera along the west coast of the United States [Master's thesis]: Athens, OH, Ohio University.

Lothringer, C.J., 1984, Geology of a Lower Ordovician Allochthon, Rancho San Marcos, Baja California, Mexico: in Frizzell, V.A., Jr., editor, Geology of the Baja California Peninsula: Pacific Section SEPM, p.17-22.

McCloy, C., 1984, Stratigraphy and Depositional History of the San Jose Del Cabo Trough, Baja California Sur, Mexico: in Frizzell, V.A., Jr., editor, Geology of the Baja California Peninsula: Pacific Section SEPM, p. 267-277.

McLean, H., 1987, K-Ar ages confirm Pliocene age for oldest Neogene marine strata near Loreto, Baja Calif. Sur, Mexico [abstr.]: Am. Assoc. of Petrol. Geol. Bull., v. 71, p. 591.

McLean, H., 1988, Reconnaissance geologic map of the Loreto and part of the San Javier quadrangles, Baja California Sur, Mexico: U.S. Geol. Survey Misc. Field Studies Map, Report No. MF-2000, 10 p., 1 sheet, geologic map 1:50,000.

McLean, H., 1989, Reconnaissance geology of a Pliocene marine embayment near Loreta, Baja California Sur, Mexico in Abbott, P.L., editor, Geologic studies in Baja California: Los Angeles, CA, Pacific Section, Society of Economic Paleontologists and Mineralogists, 63, p. 17-25.

Mina U., F., 1957, Bosquezo geologico del Territorio Sur de la Baja California: Asociacion Mexicana de Geologos Petroleros Boletin, v. 9(3-4), p. 141-269.

Minch, J.A., 1967, Stratigraphy and structure of the Tijuana-Rosarito Beach area, northwestern Baja California, Mexico: Geological Society of America Bulletin, v. 78, p. 1155-1177.

Minch, J.A., 1971, Landsliding and the Effects on Resort Development Between Tijuana and Ensenada, Baja California, Mexico: in Coastal Studies in Baja California: Office of Naval Res., Tech. Rpt. # 0-72-1, Proj. NR 387-045, 17 p.

Minch, J.A., 1972, The Late Mesozoic-Early Tertiary framework of continental sedimentation of the northern Peninsular Ranges, Baja California, Mexico: Doctoral, University of California, Riverside, Riverside, California, p. 192.

Minch, J.A., 1979, The Late Mesozoic-Early Tertiary Framework of Continental Sedimentation, Northern Peninsular Ranges, Baja California, Mexico, *in* Abbott, P.L., ed., Eocene Depositional Systems, San Diego, California, Pac. Sec., Soc. Econ. Paleontologists and Mineral., p. 43-68.

Minch, J.A., Schulte, K.C., and Hofman, G., 1970, A Middle Miocene age for the Rosarito Beach Formation in northwestern Baja California, Mexico: Geological Society of America Bulletin, v. 81, p. 3149-3153.

Minch, J.A., Ashby, J.R., Demere, T.A., and Kuper, H.T., 1984, Correlational and depositional environments of the Middle Miocene Rosarito Beach Formation of northwestern Baja California, Mexico *in* Minch, J.A., and Ashby, J.R., editors, Miocene and Cretaceous depositional environments, northwestern Baja California, Mexico; annual meeting: Pacific Sect., American Assoc. of Petrol. Geologists, 54, p. 33-46.

Minch, J.A., and Leslie, T.A., 1991, The Baja Highway, a geology and biology gield guide: John Minch & Associates, Inc., San Juan Capistrano, CA 240p.

Morris, W.J., 1966, Fossil Mammals From Baja California, New Evidence on Early Tertiary Migrations: Science, v. 153, p. 1376-1378.

Morris, W.J., 1967, Baja California - Late Cretaceous dinosaurs: Science, v. 155, p. 1539-1541.

Morris, W.J., 1969, Late Cretaceous dinosaurs from BajaCalifornia [abstr]: Geol. Soc. of Amer. Sp. Paper, 121, p. 209.

Morris, W.J., 1971, Mesozoic and Tertiary vertebrates in Baja California: National Geographic Society Research Reports, 1965, p. 195-198.

Normark, W.R., and Curray, J.R., 1968, Geology and structure of the tip of Baja California, Mexico: Geol. Soc. of America Bulletin, v. 79, p. 1589-1600.

Radamaker, K., 1995, ABA Field List of the Birds of Baja California.

Robbins, C.S., Brunn, B., and Zim, H.S., 1983, A Guide to Field Identification Birds of North America, Western Publishing Company, Inc., Racine.

Rowland, R.W., 1972, Paleontology and paleoecology of the San Diego Formation in northwestern Baja California: San Diego Soc. of Natural History Trans., v. 17, p. 25-32.

Roberts, N.C., 1989, Baja California Plant Field Guide, Natural History Publishing Company, La Jolla.

Santillan, M., and Barrera, T., 1930, Las Posibilidades petroliferas en las Costa Occidental de Baja California, entre los paralelos 30 y 32 de latitud norte: Mexico (City) Universidad p. 1-37.

Scott, S.L., 1992, National Geographic Society Field Guide to the Birds of North America, National Geographic Society, Washington D.C.

Shreve, F., and Wiggins, I.L., 1964, Vegetation and Flora of the Sonoran Desert, Stanford University Press, Stanford.

Smith, J.T., 1984, Miocene and Pliocene Marine Mollusks and Preliminary Correlations, Vizcaino Peninsula to Arroyo La Purisima, Northwestern Baja California Sur, Mexico: in Frizzell, V.A., Jr., editor, Geology of the Baja California Peninsula: Pacific Section SEPM, p. 197-218.

Standley, P.C., 1920-1926, Contributions from the United States National Herbarium: Trees and Shrubs of Mexico, volume 23, GPO, Washington D.C.

White, C.A., 1885, On new Cretaceous fossils from California: U.S. Geol. Survey Bull. 22, p. 355-373.

Wiggins, I.L., 1980, Flora of Baja California, Stanford University Press, Los Angeles.

Wilson, I.F., 1948, Buried topography, initial structure, and sedimentation in Santa Rosalia area, Baja California, Mexico: Am. Assoc. Petroleum Geologists Bull., v. 32, p. 1762-1807.

Wilson, I.F., and Rocha Moreno, V.S., 1957, Geology and mineral deposits of the Boleo copper district, Baja California, Mexico: U.S. Geol. Survey Prof. Paper 273, 134 p.

Woodford, A.O., 1928, The San Quintin volcanic field, Lower California: American Journall of Science, v. 215, 5th series, v. 15, p. 337-345.

Yeats, R.S., Minch, J.A., and Forman, J.A., 1971, Paired basement terranes in Baja California Sur, Mexico [abstr.]: Geol. Society of America Abstracts with Programs, v. 3, p. 760.

At this printing, the following, and a list of others, are available from Sunbelt Publications: 1250 Fayette St., El Cajon, CA 92020, (800) 626-6579

Adams and Wyckoff, A Golden Guide to Landforms.

Zim and Shaffer, A Golden Guide to Rocks and Minerals.

Baja Explorer Topographic Atlas Directory

Crosby, H.W., 1997, Cave Paintings of Baja California, Discovering the Great Murals of an unknown People

Automobile Club Guide to Baja California.

Gohier, F., A Pod of Grey Whales.

Gotshall, Daniel, Guide to Marine Invertebrates.

Peterson Field Guide, Pacific Coast Shells.

Peterson Field Guide, Western Birds.

Potter, Ginger, Baja Book IV.

GLOSSARY

ADAMELLITE - Coarse grained plutonic rock between granodiorite and granite in composition.

AEOLIAN - Deposited by wind.

ALLUVIAL FANS - Alluvium piled in a fan, by streams, at the foot of a mountain in a desert region. p. 6, 16:33

ALLUVIATED - Covered by alluvium.

ALLUVIUM - Sand, gravel, silt currently being moved by and deposited in streams.

AMPHIBOLITE - Metamorphic rock like a schist with aligned amphibole grains. 8:170

ANDESITE - Red-brown to gray intermediate coarse grained volcanic rock. Found in stratovolcanoes and breccias in Subduction Zones on edge of the continent. *See* Rock Chart.

ANGULAR UNCONFORMITY - Unconformity in which the lower sedimentary beds are tilted and eroded before the upper sedimentary beds are deposited.

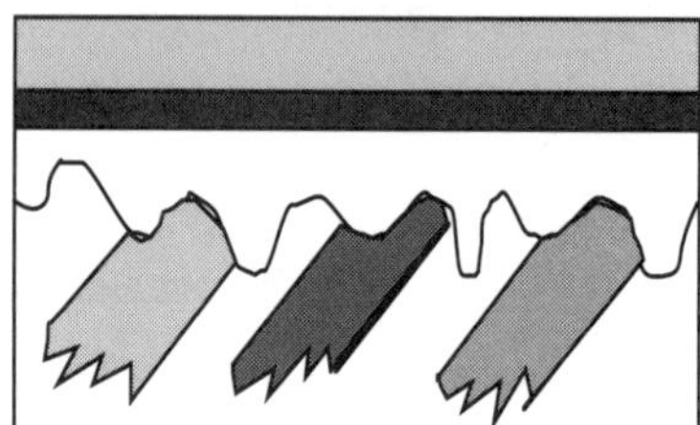

ANTICLINE - Fold where the beds are bowed up in the center

AQUIFER - Water bearing layer, usually sandstone. 2:170.2, 8:212

ARCHAEOLOGY - Study of the remains of human existence. 3:171

ARGILLITE - Massive fine-grained metamorphic equivalent of siltstone.

ARKOSIC - Sand containing abundant feldspar. Usually formed in an arid climate.

BAJADA - Coalesced (joined together) alluvial fans often surrounding a playa.

BASALT - Dark, fine grained, iron rich volcanic rock found in cinder cones and volcanic plateaus. *See* Rock Chart.

BASE LEVEL - The lowest point to which a stream can erode.

BASE LEVELED - Eroded to nearly flat plain at or near base level.

BATHOLITH - Body of plutonic rock greater than 40 mi^2

BIOLOGICAL BEACH - Beach formed from the calcareous remains of marine shells and other organisms. 6:123

BIOTURBATION - Mixing of sediments by burrowing animals.

BRECCIA - Sedimentary rock with angular fragments of rock larger than 2mm. *See* Rock Chart.

BUTTES - Small mesa or sharp flat peak.

CADODE - Stem or trunk which acts as a leaf to perform photosynthesis.

CALCAREOUS - Containing calcium carbonate.

CALICHE - Calcium Carbonate deposited just above the water table as groundwater, that is drawn upward by capilary action, evaporates.

CHEMICAL WEATHERING - Chemical break down of the minerals in a rock = decomposition; common in the tropics.

COMMENSALISTIC - A relationship in which two or more organisms live in close association and in which one may derive some benefit, but neither is harmed. 16:15.5

CONTACT - Boundary between rock units.

COQUINA - Conglomerates formed from shells.

CROSS BEDDING - Beds deposited at an angle to other beds.

DACITE - Volcanic rock between rhyolite and andesite in composition.

DEFLATION - The removal of material from a surface by wind.

DELTA - Sediments deposited in the ocean at the end of a river.

DETRITUS - Fragments of material which removed from their source by erosion.

DIKE - Igneous material filling a crack/joint in rocks. 2:89, 15:64

DIP - Angle of inclination of layers or faults. 3:112

ECOTONE - An ecological community of mixed vegetation formed by the overlapping of adjoining plant communities.

EDAPHIC - The way in which the soil affects living organisms.

EMBAYMENT - Indentation in a shoreline forming an open bay; area which contains rocks deposited in an embayment.

EMERGENCE - Area which was under water is raised above the water surface.

ENDEMIC - Native or confined to a certain region. 3:58

EPIPHYTIC - A plant that lives on another plant upon which it depends for mechanical support, but not for nutrients. 4:49

EROSION SURFACE - Relatively flat surface which has been base leveled by streams.

ESCARPMENT - Steep cliff or line of cliffs typically along a fault.

ESTUARY - Where a river empties into the ocean through a tidal area. 1:69.2, 1:73

EXUMED SURFACE - Surface buried under younger rocks and then exposed by erosion. 3:125

FACIES - Different rock units deposited at the same time in different parts of a basin.

FANGLOMERATE - Coarse sediments deposited in an alluvial fan.

FAULT - Fracture in rocks with movement.

FAULT SCARP - Uplifted cliff or bank along a fault line. 3:262.1, 11:21.3

FAULT ZONE - Most major faults are a series of smaller faults and related features. 2:45.3, 12:154, 12:127

FLUVIAL - Sediments deposited by stream action.

FOLIATED - Alignment of mineral grains.

FORMATION - Mapable rock unit.

FOSSILIFEROUS - Containing fossils.

FOSSILS - Evidence of past life. Range from original material to tracks & burrows.

FRANCISCAN - Metamorphic rocks and sediments derived from the subduction zone. First recognized in the area of San Francisco, California.

GABBRO - Dark colored, coarse grained, iron rich plutonic rock. *See* Rock Chart.

GEOSYNCLINE - Syncline on a continental scale. 5:162

GNEISS - Banded metamorphic rock with bands of dark and light minerals. *See* Rock Chart.

GRABEN - Down dropped block bounded by faults. 9:89

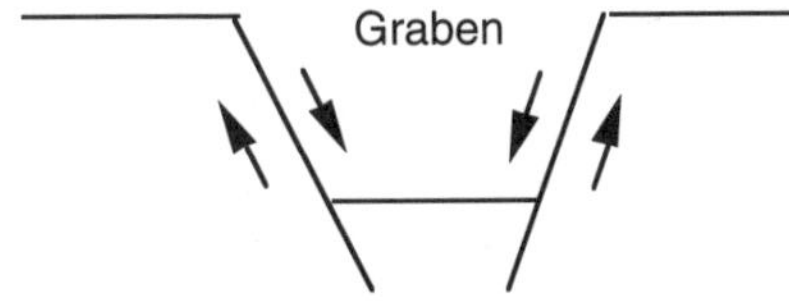

GRANITE - Light colored, coarse grained, plutonic rock. *See* Rock Chart.

GRANITIC - Used in place of plutonic.

GRANODIORITE - Intermediate composition, coarse grained, plutonic rock. *See* Rock Chart.

HALOPHYTE - Salt tolerant plants.

HOGBACK - Narrow ridge with steeply dipping beds.

HORST - Uplifted block bounded by faults. 2:151.5

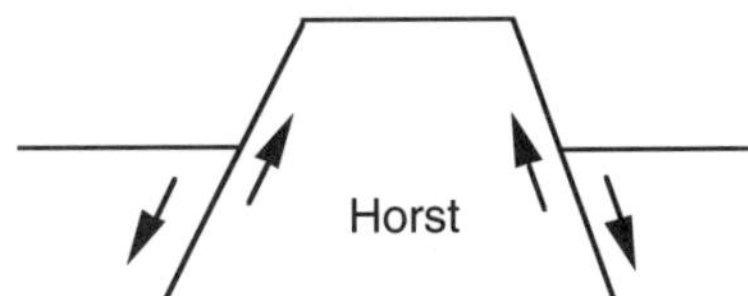

HOT SPRING - Spring in which the water temperature is 10° warmer than the mean air temperature. Usually found along faults. 2:18.5, 13:176

HUMIDITY - Amount of moisture in the air. Fog forms when the air is saturated and reaches the dew point. 7:28

HYPABYSSAL - Near surface intrusive.

INDICATOR SPECIES - A plant which is restricted to, and thus indicates a specific environment.

IGNEOUS - Formed by cooling of molten material. *See* Rock Chart.

INTRUSION - The injection of molten material into an area below the surface.

ISLAND ARC - Arc shaped group of islands formed by a subduction Zone.

ISOTOPIC AGE - Probable age determined by radiometric age dating.

JOINT - Fracture without movement.

K/AR - Potassium 40 decays to Argon 40 at a predictable rate. Heat and weathering can affect the dates obtained. (See Radiometric Age Dating)

LACUSTRINE - Sediments deposited in a lake. 3:120

LAGOON - A shallow body of water with a restricted connection to the sea.

LAHAR - Andesite flow which has broken up as it moved downslope, often mixed with water.

LANDSLIDES - General term for the downslope movement of material due to gravity.

LAPILLI - Coarse grained volcanic ejecta.

LATERAL FAULT (STRIKE SLIP FAULT) - Fault with largely horizontal motion with one side slipping past the other.

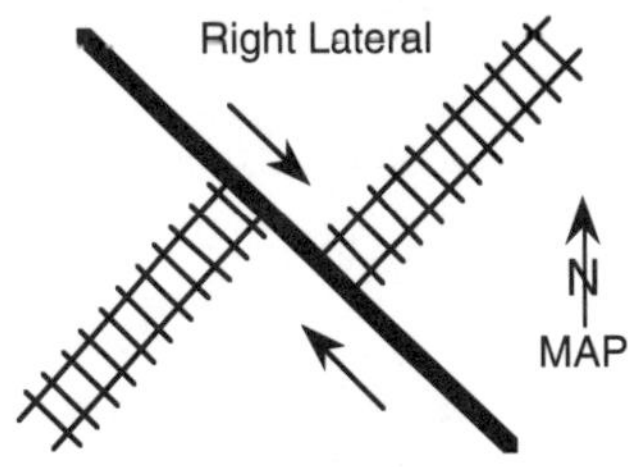

LICHENS - An algae and a fungus got together and took a lichen to each other. The algae provides the food and the fungus the shelter.

LITHIFICATION - Process by which sediment is converted to sedimentary rock (cementation, compaction, desiccation, crystallization).

LITHOLOGIES - Physical character of the rocks, rock types.

LITTORAL - Shoreline between tides.

MAAR - A crater formed by a violent explosion without igneous extrusion, commonly occupied by a small circular lake.

MAGMA - Molten rock formed by release of pressure on ocean ridges/rifts or by friction in subduction zones. *See* Rock Chart.

MASSIF - Mountain range.

MATRIX - Material around the grains in a rock.

MECHANICAL WEATHERING - Disintegration; mechanical break down - more common in deserts.

MEGA-Breccia - Large clasts in breccia.

MESIC - Moist conditions.

METAMORPHIC - Rocks which have undergone change by being subjected to heat (not melting) and pressure under the surface of the earth.

METAMORPHISM - Application of heat and pressure to change the form of minerals in a rock.

METASEDIMENTARY - Sedimentary rocks which have undergone low grade metamorphism.

METAVOLCANIC ROCKS - Volcanic rocks which have undergone low grade metamorphism. 3:112

MIXED ROCK - Area where igneous and metamorphic rocks are mixed together and not separately mapable.

MONOCLINE - Local steepening of beds, usually over a fault. 1:14.8

NONCONFORMITY - Sedimentary beds deposited on the eroded surface of igneous or metamorphic rocks. 1:52.2

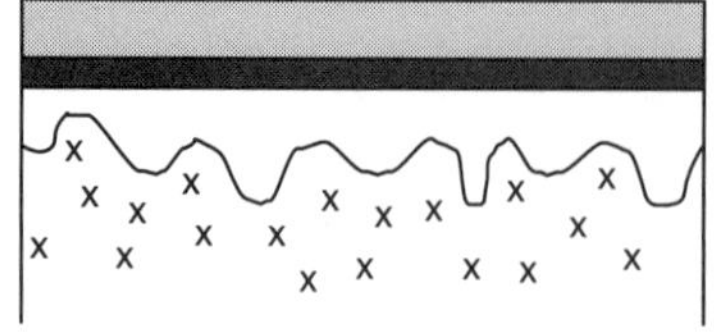

NORMAL FAULT - Caused by tension where the overhanging block slides downward. Main fault of horst and graben areas.

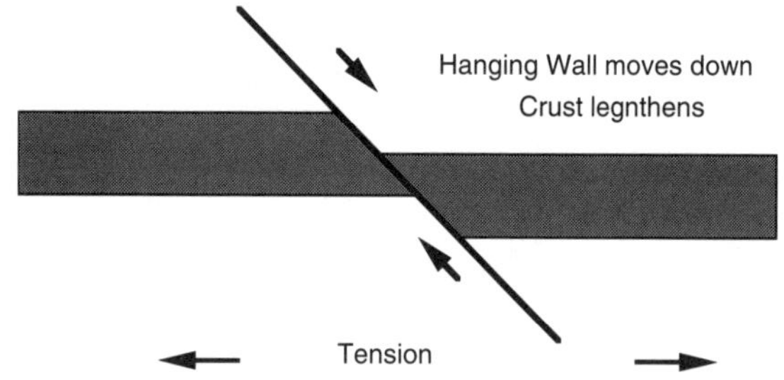

OBLIGATE - Able to survive in only one environment; usually refers to parasites.

OBSIDIAN - Volcanic glass. *See* Rock Chart.

ONYX (TRAVERTINE) - Banded calcite formed in hot spring deposits.

OPPORTUNISTIC - An organism taking advantage of opportunities for self advancement.

PALEONTOLOGY - Study of past life. Overlaps with Archaeology in past human life.

PARASITE - An organism that feeds and grows on a host, eventually harming or killing it. 4:49

PEDIMENT - Gently sloping surface cut by streams, in an arid climate, at the foot of a mountain range. Grades into an alluvial fan.

PEGMATITE - Coarse grained dike, often with rare minerals such as beryl, tourmaline, or topaz.

PENEPLAIN - Low relief erosion surface cut by streams in a humid climate.

PHYLLITE - Foliated metamorphic rock with subvisible mica.

PHYTOGEOGRAPHIC - Groupings of plants by geographic occurrence.

PLATEAU - Extensive area of volcanic flows, usually basalt, with occasional cinder cones.

PLAYA /PLAYA LAKE - Desert area of interior drainage wherewater and fine sediments accumulate. Usually dry.

PLUGS - Eroded necks of volcanoes. 1:39, 13:166.5, 16:15.5

PLUTON - Individual body of plutonic rock.

PLUTONIC - Coarse grained igneous rock which cooled slowly deep (miles) in the earth. *See* Rock Chart.

PORPHYRY - Igneous rock with larger crystals in a finer grained matrix. *See* Rock Chart.

PREBATHOLITHIC - Rocks formed prior to the intrusion of the Batholith.

PTYGMITIC - Tightly folded in serpentine manner.

PUMICE - Volcanic froth from gas filled magma. *See* Rock Chart.

PYROCLASTIC - Fragmented volcanic material often ejected into the air from a volcano.

OUARTZ DIORITE - Roughly equivalent to tonalite.

RADIOMETRIC AGE DATING - Certain elements are radioactive. That is, they change from a parent element to a daughter element. This change is random. However, the rate is predictable with large numbers. This rate is called a half life which is the time for 1/2 of the parent element to decay to the daughter element; 1, 1/2, 1/4, 1/8, 1/16. The ratio determines the age.

REVERSE FAULT / THRUST FAULT - Hanging wall block is pushed up and over the footwall block. Formed by compression.

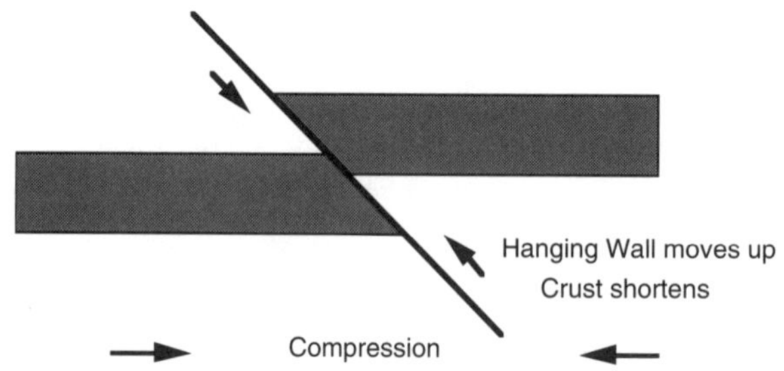

SCARIFICATION - Abrasion of the coating on a seed by running water.

SCHIST - Foliated metamorphic rock with visible aligned minerals such as mica and amphibole. *See* Rock Chart.

SEDIMENTS - Fragments of preexisting rocks worn from the land and deposited by wind, water, ice, or gravity.

SEDIMENTARY - Rocks made up of fragments of rocks deposited by running water, wind, ice, and gravity, or by accumulation of various organic materials formed by chemical action.

SHUTTER RIDGES - Formed when a fault moves a ridge to block the drainage of a low area.

SILL - Tabular intrusive body which is parallel to layering.

SLUMP - Rotational block slide.1:62.3

SPHEROIDAL WEATHERING - Igneous rocks have a tendency to weather into spheres since the elements attack corners three times, and edges twice, as fast as rock faces. 3:157

SPIT - Strip of sand or mud, deposited by waves or currents, projecting from the shoreline. 6:184

STACK - Small isolated, near shore rock mass. 9:6

STRATA - Layering in sedimentary rocks due to changes in regimen.

STRATIGRAPHIC - Pertaining to the description of the strata.

STRIKE - Horizontal bearing of tilted layers or fault line.

STRIKE SLIP FAULT - A fault which has lateral movement.

SUBDUCTION ZONE - Where one plate dives beneath another along a convergent plate boundary.

SUBMARINE FAN - Sediment moving down a submarine canyon forms a fan shaped mass at the end of the canyon.

SUBSTRATE - Rocks or sediments at the land/water interface.

SYNCLINE - Downfold or basin where beds drop towards each other.

Beds are folded towards center

TALUS - Landslide where the movement consists of a few rocks at a time.

TOMBOLO - Strip of sand or mud, deposited by waves or currents, connecting an island with the mainland. 6:93.2

TONALITE - Intermediate composition, coarse grained, plutonic rock. *See* Rock Chart.

TUFF - Volcanic fragments, usually ejected from volcano.

TUFFACEOUS - Containing tuff

TURBIDITE - A turbid mass of muddy water moving down the continental slope.

UNCONFORMITY - Gap in the rock record usually represented by a period of erosion.

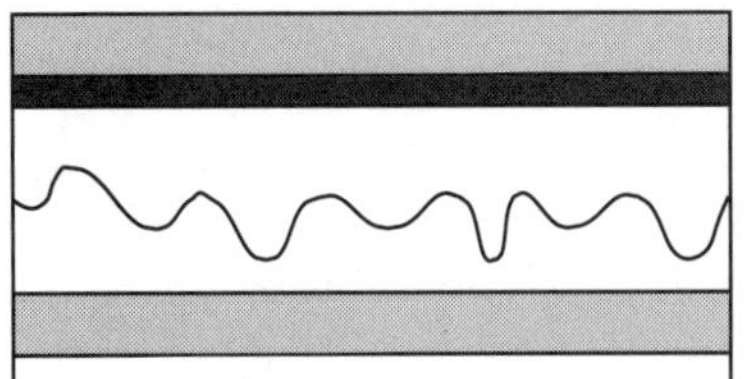

Deposition of upper layers

Uplift & Erosion

Deposition of lower layers

VOLCANIC - Fine grained igneous rock which cooled rapidly at or near the surface. *See* Rock Chart.

VOLCANICLASTIC - Fragments of volcanic rock.

Xenolith - Partially melted remnant of older rocks included in an igneous rock. Usually darker in color. 9:2

XERIC - Dry conditions.

XEROPHYTE - Plants adapted to dry conditions. 3:84

BAJA CALIFORNIA GEOLOGIC TIME SCALE

ERAS	PERIODS	EPOCHS	EVENTS IN BAJA CALIFORNIA
Cenozoic	Quaternary	Holocene	Continued faulting & uplift
		Pleistocene	Principal uplift and tilting of Ranges
	Tertiary	Pliocene	Mouth of Gulf of California opens
		Miocene	Widespread volcanism-central Gulf opens
		Oligocene	First movements on lateral faults
		Eocene	Major, gold bearing, rivers flow across area
65my		Paleocene	Shallow coastal seas and tropical weathering
Mesozoic	Cretaceous		Subduction Zone, Formation of Batholith,
	Jurassic		Cretaceous Geosyncline, and Coast Ranges.
230 my	Triassic		Time of Dinosaurs in Baja
Paleozoic	Permian		Volcanism
	Pennsylvanian		shallow seas ? over much of
	Mississippian		Baja California
	Devonian		
	Silurian		
	Ordovician		Oldest rocks in Baja ?
600 my	Cambrian		
Precambrian			

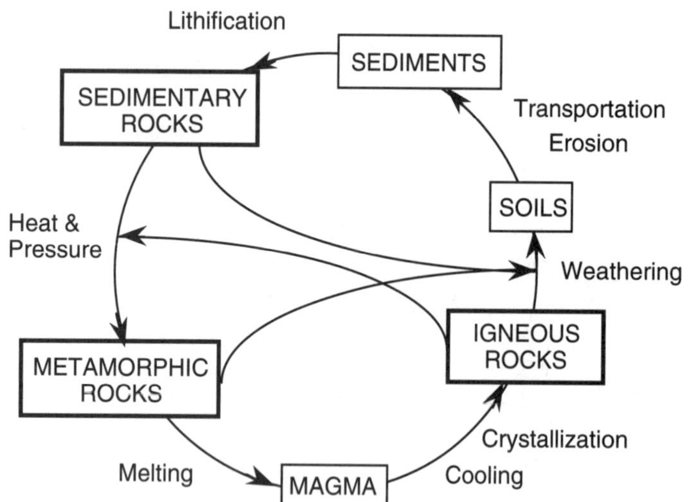

ROCK CYCLE

All rocks are inter-related and part of the cycle of the earth and will eventually go through the complete cycle to molten material. Once rocks are on the surface they tend to stay there and be recycled through the weathering process and erosion.

MAGMA cools and crystallizes to form IGNEOUS ROCKS which weather to form SOILS, etc.

IGNEOUS ROCKS

	Silica Rich		Iron Rich	
Plutonic	GRANITE	GRANODIORITE TONALITE	GABBRO	Coarse Grained- Cooled slowly deep in earth
				↔ Visible Limit
Volcanic	RHYOLITE	ANDESITE	BASALT	Fine Grained Cooled rapidly at or near the surface
	Light		Dark	

Igneous Rocks are formed by crystallization from a magma (molten rock). They either cool slowly, deep in the earth, and are coarse grained. Or, they cool rapidly at the surface and are fine grained. Some (porphyries) cool slowly and then rapidly forming both large and small crystals

OBSIDIAN = volcanic glass
PUMICE =frothy volcanic glass
TUFF = fragments from explosion

	Sediment	Sedimentary Rock	
Clastic (Fragments)	Gravel	Breccia △ Conglomerate ○	
			2mm
	Sand	Sandstone	
			1/16mm
	Silt/Clay	Siltstone Shale	
Nonclastic (Chemical)		Chert [SiO_2] Limestone [$CaCO_3$] Gypsum [$CaSO_4$]	

SEDIMENTARY ROCKS

Sediments are formed by weathering and erosion [by water, ice, wind, gravity] of rocks on the earth's surface. Lithification of sediments by cementation, compaction, dessication [drying], and crystallization produces Sedimentary Rocks

BRECCIAS are angular and CONGLOMERATES are rounded fragments larger than 2mm

METAMORPHIC ROCKS

SLATE → SCHIST → GNEISS
Limestone → MARBLE
Sandstone → QUARTZITE
Mantle rock → SERPENTINE
Volcanic rocks → METAVOLCANIC ROCKS

METAMORPHIC ROCKS are formed by the application of heat and pressure to other rocks. This occurs below the surface of the earth.

Pressure compresses SHALE to SLATE. More heat and pressure causes mica to grow forming SCHIST. As feldspars grow the GNEISS with banding is formed. LIMESTONE becomes MARBLE and SANDSTONE becomes QUARTZITE. SERPENTINE is considered altered mantle material.

METAVOLCANIC ROCKS are metamorphosed volcanic rocks.